Social Media Myths...
BUSTED

Social Media Myths...
BUSTED

The Small Business Guide To Online Revenue

by Laura Rubinstein, CHt

Social Media Myths Busted:
The Small Business Guide to Online Revenue

Cover design by Bertz Santos and Laura Rubinstein.

Edited by Sheri Horn Hasan,
Karmic Evolution Editorial & Publishing Services LLC,
www.KarmicEvolution.com

Book design by Maureen Cutajar
www.gopublished.com

Printed in the United States of America
First Printing, 2014

ISBN: 978-0-9749845-7-5

Transform Today
PO Box 34105
San Diego, CA 92163
coach@TransformToday.com
www.TransformToday.com
www.SocialMediaMythsBusted.com

For inquiries about author appearances, ordering in quantity, or permission to reproduce, please contact the publisher at the address above or contact Transform Today at Tel: (619) 940-6569 or visit www.TransformToday.com.

DEDICATION

This book is dedicated to the small business owner and his/her entrepreneurial spirit. For it's the small businesses of the world that are making the biggest and most important impact on people's lives. Large corporations have shareholders to make happy. A small business is directly accountable to the end customer. Thus, the work of a small business owner—one with immense passion and deep commitment to customer delight—ultimately makes the world a better place.

CONTENTS

FOREWORD

I t's a day just like any other day. You gather your mail from your mailbox and, while sorting through the junk mail and bills, you discover a mysterious envelope. Hand-addressed in calligraphy and lacking a return address, it reveals upon opening an invitation.

Apparently, you've been invited to a party of some sort, but the sender remains a mystery.

Curiosity gets the best of you and you show up at the event. Hundreds, perhaps thousands, of people are scattered throughout the massive ballroom. Groups of two, three, four, and five or more form casual circles, and both beverages and discussions flow freely.

You eavesdrop on conversations taking place in real time as you walk past these groups. One vigorously debates current events. Another discusses the previous evening's airing of a popular television drama. Two people debate which New York football team holds the greater legacy, while another group laughs hysterically as they show photos of who-knows-what to each other.

But it's the group of five discussing your particular hobby that gets your attention. When you overhear one person in this circle ask a question, you interject excitedly because you know the answer! Your contribution to the conversation is rewarded with a warm welcome from the group. You think you'll hang out here for a while and get to know the others in this group.

Welcome to social media.

When you strip away the face-to-face interaction that takes place at an event as described above, what you're left with is the content of engagement with other human beings...the social aspects of communication.

Some people think of social media as a different form of communication. But it's not. This is just one of many myths people believe about social media.

You see, social media is engaging with others, only from remote locations. Whether at your home or office computer or on your mobile phone or tablet, social media is wherever *you* are.

Whether you're seeking to leverage the power of social media for business or pleasure, gain a greater understanding of what *is* social media exactly, and/or how you can maximize it, you'll be well-served by reading this book.

Laura Rubinstein *knows* social media. As one of the founders of *Social Buzz Club,* she has a lot to teach you as you read through these pages.

Not only will you learn a lot about the giving and receiving of relationship with others through this suddenly ubiquitous form of interaction, but as your knowledge grows you'll become more proficient in your ability to engage your business and enhance your brand in the social space.

And in this, I believe, you will find the freedom to be a more authentic you.

—Joel Comm, author of New York Times Best-Seller
*Twitter Power: How to Dominate Your Market
One Tweet at a Time*

INTRODUCTION

You may be thinking *is this another one of those social media books that will tell me social media is great by showing how household brand names and multimillion dollar corporations use these networks?*

If you're like me, you cannot relate to the big marketing budgets, availability of substantial staff resources, and media attention these large brands garner.

Thus, you're wondering how in the world you, the small business owner, can get any traction on these noisy systems. Perhaps you're overwhelmed by the complexity of social networking platforms. Or maybe you simply feel resistance to putting time into developing a strategy.

In fact, you're probably wondering *is social media even worth it at all?*

The short answer is *yes!*

I set out to discover what kinds of common "myths" are being circulated. By speaking with a group of successful small business owners, as well as social media pros, I identified the TRUTH behind these myths.

I also identified the ways people are succeeding at making

money on the top four major social media networks, where I found that *real* revenue *is* being generated!

Welcome to my book, where I make these secrets and more based on my own expertise available to you now!

What I've found after years of immersion into the social media realm is that the primary block to making social media work comes from *our way of thinking*. Social media has been a huge trend since the early 2000's, but in the beginning many naysayers were sure it was just a fad.

As a result, huge resistance surfaced. As time went on, the naysayers' predictions about social media's quick demise morphed into complaints that it takes too much time, that there's no return on investment (ROI) or profits, and that it's just a bunch of people sharing what they ate for lunch. These complaints lingered and infiltrated the minds of innocent bystanders.

Myth: A widely held but false belief or idea

Perhaps you're one of these innocents who's been impacted by myths perpetuated by these naysayers. You *are* if you recognize this feeling in your body: the tightness that comes about when you ponder the learning curve, your concern about negative exposure or the expense of hiring someone to run your social media.

I can relate. In the early 2000's people sent me connection requests on LinkedIn. I reluctantly accepted. I didn't believe it would catch on so I didn't think much about it. The next time I did think about it, there were millions of people connecting and doing business because of these interactive networking sites.

Boy, did that wake me up! A marketer for more than 20 years, I couldn't deny that having a place to connect easily with potential clients of all kinds is a powerful asset for the small business owner.

In this book, I've interviewed many of my social media expert colleagues to get their take on the social media myths that have persisted. You'll be interested to know that many of them had to overcome initial resistance to social media. The key take away is that once you make new mental connections, adopting this new communications system becomes natural and profitable.

One great example is Sierra Modro, technology evangelist and former Mobile Technology Evangelist for Intel, revealed that her first impression of Twitter was that it was one of those "this is what I had for lunch" sites. She didn't see a whole lot of purpose for it. Nor did she see how Facebook was going to be relevant to her work. It wasn't until her fellow master's program students started suggesting she use Facebook and Twitter that she became open to taking it seriously.

Modro's first "aha" moment occurred when she connected the dots from her technology experience to social media—only then did it begin to make sense. She realized that social media was the new version of a UNIX command she learned in 1989 named "talk."

Modro used "talk" to chat with co-workers and others by typing on her computer system at work. She recognized that Facebook, Twitter and even LinkedIn were new ways of connecting with people online.

As a result, Modro increased her influence, connections, and opportunities substantially.

"I created one relationship that lead to another that creates another. And suddenly I've got all these really interesting people who are resources for me and for whom I'm a resource," she notes. In fact, Modro and I met because of a connection she made through Twitter. Since then, she's been an incredible resource to my business.

"All computing is social." ~ Sierra Modro

Modro and I are typical examples of two people who initially resisted social media and then had a change of heart and mind. You'll find many others cited in this book—and probably in your own life—who started out reluctant to use social media and who have now become the biggest advocates of its platforms.

In my consulting practice I work with passionate business owners, guiding them to optimize and streamline their social media presence, to raise awareness of their brand, and to attract new opportunities and clients. My mission is to fuel their business and provide them with the necessary keys to become well respected and widely known "go to" leaders in their field using the power of online networking.

Thus, I spend time keeping up with changes, updating strategies, and communicating with colleagues, prospects, and clients, in order to craft the simple steps that allow my clients to thrive online.

For me, it's interesting and very creative work. As a social media strategist and marketing consultant, my job is to stay current and to distill social media marketing down to its most effective strategies—the ones that will get my clients' social media optimized and make it work for them. I do the heavy lifting, so to speak...

"I spend time staying updated on social media so others don't have to." ~ Laura Rubinstein

If you find yourself complaining about social media and thus avoiding it, you're missing out on one of this century's finest opportunities for the small business owner to connect deeply to their marketplace. The more willing and enthusiastic the user, the more serendipitous and fruitful their experience.

From the earliest days of modern social media—when professionals began receiving requests to join LinkedIn—they started to find business opportunities they would otherwise have missed had they not reconnected with an old colleague or accepted a connection with a new one.

Next, individuals began to connect with alumni on LinkedIn and Facebook and, as a natural result of these connections, more business wound up happening. These seemingly small random connections actually generated a whole lot of revenue. Now we see both small business as well as large jumping all over social media advertising. Later on, we will cover how the small business can take advantage of the advertising features.

When I speak to groups and organizations made up of business owners, often I ask "how many of you have gotten business from social media?" Maybe 2% of hands in the room go up. Then I ask "how many of you have done business because you reconnected with someone from your past on social media?" Invariably 40-50% of hands are raised.

The truth of the matter is that many people have gained business as a result of social media. This one little shift in perspective about earning revenue through social media is eye-opening for many reasons.

First and most importantly, making social media profitable is super easy—you simply have to believe that it's completely doable. Next, by following some easy practices—revealed in this book—you, too, will naturally increase your results online.

As a business owner in today's social media environment you can't afford to just give it a try and hope it will all work out. You need to know what works and how to most efficiently grow your visibility as an industry leader, authority, expert, or up-and-coming brand name. With a little consistency, you'll notice how your confusion and overwhelm disappears and becomes a thing of the past.

In this book I present excerpts from my interviews with successful business people and owners, who describe some great tactics and help set appropriate and realistic expectations.

The second part of this book is organized by social network. I provide you step-by-step action guides that you can follow easily. Put in an hour a day and you can garner results in as little as a month!

The main networks I cover include Facebook, Twitter, LinkedIn, and Google Plus. These are the most popular networks that offer credibility to brands. In other words, if you're not there, you run the risk of losing potential customers who can't find you.

Other networks like Pinterest and Instagram are important to keep in mind as well because they're powerful for certain niche industries (for example: fashion, food, fitness, beauty, wellness, personal development, and business development.) Although I have not designated a separate section for these, I offer insights about these networks from time to time to enhance your results.

Get ready to learn small business strategies that bring in a huge number of leads, qualified prospective clients, new business relationships, opportunities, unprecedented joint venture connections, and more!

HOW TO USE THIS BOOK

This book not only presents the myths and misconceptions I hear consistently from small business owners, thought leaders, and entrepreneurs, it walks you through a "myth-busting" process and turns these myths into useful strategies for business growth.

Until you gain clarity about these myths, you may still be resistant to—or afraid of—using social media. Such resistance is the ultimate *saboteur* of your online success! This is especially true if the only thing about which you're clear is that you are confused or fearful. If so, you'll remain paralyzed from using these networks to your greatest business advantage.

My goal is to guide you to breakthrough your resistance, and to transport you onto the road to revenue. I will present these business-stalling myths, *bust* them, and offer you guidance on how to get moving! Then I will ask the questions that help you create your social media success plan!

This book presents 4 major *myths* about online social media marketing:

1. Myths That Keep You Broke
2. Myths That Keep You Stuck

3. Myths That Put You At Risk
4. Myths About Specific Social Networks

You'll learn how marketing professionals think strategically and grow their businesses using social media. Then you will receive coaching guidance to develop your social media profit strategies. Each section covers a variety of myths, and each myth has its own chapter that:

- Clearly identifies the **Myth**
- Reveals how it's **Busted**
- Shows you **How To Bust Through** your resistance
- Offers **Coach Laura's Breakthrough Questions**

My breakthrough questions are designed to help you understand what you need in order to develop your own strategic social media marketing plan that builds revenue. As a business coach for more than twenty years, I know that asking questions is far more important than giving you answers.

By the time you reach Section 4, Myths About Specific Social Networks you'll be ready to step into each network and instead of breakthrough questions, I offer you an Action Plan for each network.

After all, I *know* you already have the answers! But in order for you to realize you do, you need some well-crafted questions! You know your business best—and when you ask yourself the questions within each section, your next steps will be clear.

SECTION 1

SOCIAL MEDIA MYTHS
THAT KEEP YOU BROKE

Every day you'll hear a certain part of the business population complain about social media. Perhaps you have a few complaints yourself! Regardless of these complaints, social media continues to be a trusted source for consumers to find out about a brand's reputation, learn of special offers, and receive good customer service.

Think about the reviews you read on Google, Angie's List or Yelp. Or if your friends on Facebook had a great meal, bought a new gadget or had a great experience at a new service provider. That third party unsolicited endorsement is HUGE and very meaningful to consumers.

The media also relies on social networking as a source of news. Many journalists scout news stories on Twitter, Facebook, Instagram, LinkedIn, Pinterest, and Google Plus. If you have complaints about social media, you're missing out on potential sales and the development of customer awareness of your brand.

The continuous perpetuation of the following myths thwarts many a small business's success.

Myth: Social Media Sites Are Purely Marketing Platforms

When a business decides to establish its presence on social media networks, often it tries to implement mass media marketing strategies. The problem is that no one likes to be marketed to *anymore.*

"It still boggles my mind how people sign up for Twitter and/or Facebook, and immediately start sharing about their product or service, as though you'd simply move in, walk down your neighborhood street, knock on the doors of your neighbors, and tell people what you sell—all without getting to know them at all!" says web pioneer and social entrepreneur Joel Comm.

"Without allowing them to get to know you it makes absolutely no sense" adds Comm, author of the New York Times Bestseller *Twitter Power: How to Dominate Your Market One Tweet at a Time.*

"It short-circuits the 'like me, know me, trust me, pay me' process by jumping right to the end of it. Of course it doesn't work because no relationship has been built there."

Busted

You may be thinking: *How can I benefit from social media if traditional marketing strategies don't work?*

Say you establish a Facebook business page and simply feature your vitamin supplement products there. In today's world of social media marketing, this isn't going to work. So you add the image of a beautiful woman along with your vitamin supplements. Still, that isn't going to garner much interest.

If you think about social media as a cocktail party or giant networking gathering, you realize the interactive nature of this virtual world. In fact, social networking has changed the marketing paradigm because of the various two-way communication features.

Think about it like shaking hands—a two way exchange with real people. On social media, some form of engagement or acknowledgement is the equivalent of a hand shake. You extend your hand by asking a question, offering a valuable app, a free report, or simply by responding to a comment from someone who's reached out to you. This makes your brand more human.

"Social media is about relationships—it's about engaging and interacting,'" agrees Ted Rubin, the successful marketing strategist of Brand Innovators who is known for evangelizing the term "ROR," or "return on relationship."

"It's an opportunity to go well beyond advertising and marketing's traditional format and takes us back to a time when we really got into relationships with customers," Rubin notes.

The beauty of social media is that *you*, the small business owner/entrepreneur, have the opportunity to be seen if you engage, interact and build tight communities with your market.

Consumers *want* that connection to their product and service providers. Too many reports of consumers being taken advantage of, paying a high price for low value, and companies that are impossible to contact, have left potential customers craving better service and attention.

We hear all too often about businesses failing to act in the consumers' best interest and actually causing harm—or in some cases death! Nowadays, in fact, consumers are so leery of corporations they expect companies *not* to act in the consumers' best interest. Although that sounds bleak and rather pessimistic, think of this as *good news* for *your* emerging business.

You have a HUGE opportunity to position yourself as the consumers' best friend and to build advocates using social media. It's the perfect tool for bringing back into the economy trusted relationships with businesses.

"When you are marketing right, you are building relationships," explains Comm. "You don't have to promote so much. Instead, people are automatically drawn to those they know, like, and trust. That means they've demonstrated they care about other people and that they bring value to the conversation—whatever it is.

"It's why we refer. It's why I like to refer the guy from whom I just bought a car. In fact, this mechanic here in Denver who could have totally taken me for parts and labor, but instead told me to junk my car and get a new one, is getting consistent referrals from me. I'm telling everyone about Saul in Denver. It's common sense...

"There's a major problem with social media, and it's this: it is not commerce friendly," Comm continues. "There's no way for brands to truly reach customers who *want* to hear from them without spending money. That's the Achilles heel of all the social sites. It really paves the way—it opens the door—for a new site to come in that will support both consumers and merchants.

"This isn't rocket science, it's not brain surgery, it's not nuclear physics," concludes Comm.

How to Bust Through

"Just interact, engage, and look for people who are discussing topics that are of interest to you," Comm suggests. "Read what they say. Then follow those people that interest you. Engage with them. Leave comments, like their posts. Share what they're saying and then bring value to the conversation so that they'll realize, 'this person has something to say—I want to follow them back!'

"This is how you build your tribe," he explains. "You don't build it by going out there and asking people to follow you. Rather, you bring value, and when you do people will automatically want to engage with you because you're engaging. Be engaging, be interesting, and be interested in knowing what makes your market tick.

"If you're a thought leader, tell stories that inspire, share problems and solutions, ask for input," Comm advises.

Coach Laura's Breakthrough Questions

- Do you understand who your market is?
- Do you know where they hang out on social media?
- In what are they most interested—what engages them most?
- Are you clear about your social brand message and does it match your market?
- What does your target market and audience *do* on social media?
- What is the craving that you can fulfill for your market?

Myth: Promoting My Business On Social Media Will Drive Customers To My Door Instantly

Once a new business becomes aware of the fact that they need to get going on their social media presence in a more focused way, often this creates high expectations. They believe that—the moment they put up their business accounts on any given network—millions of people will see it and their phone will start ringing off the hook.

While some realize they need to get set up right, they still think the same thing: that having a Facebook or Google Plus page means customers will instantly arrive and bang on their door.

Busted

Did you know that when you put up a brand new Facebook business page, almost no one will see it? *Why is that?* Think of it like placing a 8.5" x 11" flyer on the wall of a baseball stadium. It's there for everyone to see, but it's too small for anyone in the stands to notice. Even a player who walks by it on his way up to bat or back to the dugout isn't likely to pay any attention to it!

In other words, the phrase "if you build it, they will come" is a myth on social media if you don't build it with strategic visibility, engagement, and a good marketing plan. Additionally, several things have to be in place before the masses will begin to arrive. Finally, in reality you don't want the masses, you want targeted individuals or groups who want what you have to give them.

Whether you love social media or hate it, it's a very noisy place. When you set up your social media accounts, though they are public, no one knows about them. It's like having an advertisement in the yellow pages. People have to know exactly what to look up to find you, and then sift through other competitor ads until they find yours.

Then they have to *choose* to call you. Similarly, on social media if you are one of several thousand authors, wellness companies, artists, coaches, or other service professionals, your social media listing is not going to stand out on its own.

"You can't make yourself influential. You can only gift influence to other people and they can give it back to you. Just like if you want to get respect you have to give respect. If you want to get influence, you want to give influence," Teresa de Grosbois, influence mentor and founder of the Evolutionary Business Council, points out.

"Social media is made for people and is about relationships," explains Neal Shaffer, social media strategy consultant and founder of the blog *Maximize Social Business*. "It takes time to get to know people.

"Social media is no different...it takes time, and you're not going to get results overnight" explains Shaffer, the author of *Maximize Your Social,* and two other books about LinkedIn. "Some are lucky and go from here to there *because they built out that platform—there's this pent up demand* for engagement, and they come in with the right piece of content at the right time and *boom!* You know, the 'will it blend?' example.

"But it takes time and a lot of planning and resources to get there," he adds. "Do not be fooled—it does not happen overnight! Take your time and realize that having a social media presence is like having a website. It becomes part of your infrastructure over time."

Social media is so much more powerful when you include it as part of a comprehensive marketing plan. Deb Cole—a consultant with *New Media* and author of *Twitter Revolution*—does this really well for herself and her clients.

"I'm not just on social media" Cole explains. "When consulting with clients, I look at everything marketing related. And if people don't go to expos, conferences, and events to meet people face to face, you lose the power of what you get in social networks.

"When we use social networks, we get to know people before an event and continue to nourish these relationships in real life (IRL.) We meet people at an event, and that's when they get to bond with, know, like, and trust you. That's when business comes quicker because they see you top of mind on social networks. They keep seeing you on social media, and they'll come back to you when they're ready.

"I'm not here and gone," Cole adds. "I'm here for the long haul. Marketing becomes much simpler when you adopt social media and bust the myths out there that your competitors are holding on to."

How To Bust Through

To stand out on social media you need a multi-pronged and well organized approach. The system for building a strong social media foundational strategy includes optimizing profiles, crafting compelling content, developing ongoing engagement, and instituting a tribe-building strategy.

Optimize your profiles with your website link(s). Not making a website easy to find is one of the biggest mistakes social media marketers make. Ensure your website link is placed not only in the website field of each of your profiles and pages, but in summaries, "about" sections, and on cover images as well.

Optimizing profiles and posting compelling content doesn't guarantee sales. Driving sales comes through the strategy development and relationship building process, and includes integrating a strong plan that combines your website, online engagement, and value offerings such as webinars, downloadable checklists, white papers, and hot topic reports.

This integration must be systematized and lead individuals down a path they enjoy every step of the way. In the end, it should be a no-brainer to buy from you! At some point, you may even need to take the connection offline to an event or phone conversation.

A great website will guide visitors through a process that builds a relationship and leads to a purchase. You may need to re-think your website so that you attract potential clients/consumers into what is known as a "marketing funnel." Hint: This is where an email marketing system comes into play. Through such a system, you guide them to understand what you have to offer and its value.

Perhaps you provide a sample, and explain how they can get more from you in a series of email messages that complement the information on your website. Although this book is not about building a marketing funnel, it is an essential piece in your social selling success puzzle.

Once your website and marketing funnel are in place, you're ready to plan your content and engagement. Content is the proof that you have valuable, relevant, and rewarding information for your market. It's the way people know your products and services are of the highest quality—because your available free content is so valuable.

Think about content in terms of media as well. Consider—as part of your content repertoire—images, video, text, links to blog posts, and resources other than your own (links to others' articles.)

This is where you may want to get creative. Think about how you can stand out. When you develop your brand, think about your brand personality. Is it funny or sarcastic? Is it like Mary Poppins or a rap star? Make sure whatever personality style you choose appeals to your profit market and allows you to communicate through your *authentic voice.*

I like to think of this as a theme. For example, my Facebook profile theme is *inspiration.* So, even though it's personal, it still has appeal to my market. Often you'll see here images with quotes, and questions meant to inspire new ideas and ways of thinking about life, communication, and business. Sometimes my posts are solely for human interaction and to show my personal side. This is important, too. If your social media seems too impersonal, it will not gain any traction.

Engagement shows people they matter, that your company has a human face, and that you not only care about sales but about *them.* This means that someone must check in on your social network accounts and respond to comments, mentions, tags, and messages. Be sure to respond *regularly.*

What is regularly? I suggest starting with at least three times/week but, ideally twice a day. If no one is engaging with you, then I suggest proactive engagement with others. That would start with the basics. Reach out and add colleagues, existing clients, and business acquaintances as friends on Facebook, and follow them on Twitter. Thank them when they accept your friend request or follow you back. You might even comment on their recent posts.

If you use Facebook as your page, then you can go to other industry expert pages and comment on some of their most popular recent posts. Simply add to the conversation. Be mindful not to promote yourself. Rather, contribute a thoughtful positive piece of advice or encouragement. You might also want to congratulate your LinkedIn connections on their new positions, like their updates, and/or invite new connections that LinkedIn recommends.

Ideally, over time you will create a tribe of people who are your colleagues and raving fans who love your content and contribution that you make in through your products, services and message, and they naturally share your compelling content. Many influencers use content amplification systems like the one I co-founded along with Kathryn Rose called the Social Buzz Club. In this online platform you can submit your content for members to curate to their social networks. You might be thinking, *why would others want to do this?*

Adding valuable content from other sources that you can comment makes your timeline much more interesting. Offering a wide variety of content makes you a valuable resource to your audience. You are also giving influence. The more you give influence the more you will receive it back either directly or indirectly.

You may want to set up Google Alerts, or choose from a variety of monitoring tools to find out who's chatting about your company or your staff, and to keep abreast of what news is out there about you. Google Alerts is also great for informing you about competitors' newly published news and industry trends.

Many business owners worry unnecessarily that a negative comment will damage their reputation. Let me settle this myth for you very simply: you are at risk of that every day just by being in business! Someone can easily report you to the Better Business Bureau, or tell a slew of friends via email, or conversation, about you. However, if they comment on social media you have a wonderful opportunity to turn that customer around.

Having been a customer service manager for a number of years, I know that even the most hateful of customers can change their attitude when they are reached out to. You can do this publically on social media and let them know you are there, willing to help. You can offer them a way to be in touch.

When the public sees you taking action to set things right, you increase your integrity in the marketplace. If customer service is lacking in your business, I highly encourage you to revamp your system and ensure that customers are satisfied more often than not. If you can't, social media will most likely be a place that will bring the truth to light in a bigger way.

Social media is the icing on the marketing cake. It is *not the cake*. The cake must include other forms of marketing. If you're a coach, consultant, therapist, or professional service provider, a good mix of relationship marketing strategies will also serve your business well. These include encouraging referrals, public speaking, networking offline, and being well connected in your industry.

Once you have a complete *cake* (i.e., marketing plan), a beautiful thing can occur, and you'll be able to nurture and develop a robust tribe of leaders, customers, prospects, and brand advocates.

Coach Laura's Breakthrough Questions

- What is the path from relationship to revenue through which you want to take your social connections?
- What content will provide potential customers the kind of value they're willing to pay for *and* garner great feelings about your brand?
- When will you optimize your social accounts?
- Who are the influencers you want to build up and be seen associating with?
- For which topics, companies, and people will you set up Google Alerts?
- Who will be responsible for online engagement and at what frequency?
- How will they engage and build your tribe?

Myth: I Should Build My Platform On Social Media And Keep It There

Some small businesses rush to social media and spend a tremendous amount of time communicating there. They may even spend a lot of money advertising to get Facebook "likes" and Twitter Followers. They may make hundreds or thousands of connections on LinkedIn and build their Google Plus circles into the tens of thousands.

In the meantime, their website doesn't have any links to their social media, it's not responsive (meaning they failed to design it for viewing on a mobile device), and they don't have an email system or marketing funnel. All they have is a large platform on social media.

It's nice to grow a large following on the social networks, but how is this helping your business grow?

Busted

It *is* possible to build a business on social media *only* and without a website. I've seen it done successfully. However, in my opinion it's a bit risky and you run the risk of losing everything.

For example, suppose you've accumulated 50,000 connections between your top three networks, and one day your best, most engaged network changes its rules or determines you've violated their terms. They could start making your business pay for every interaction or to share links. They could remove your connections because of some algorithm change or, worse yet, they could close your account and never let you have access again with no good explanation.

By the way, this is not just hypothetical speak! I've had my LinkedIn account shut down because I used a feature they encouraged, and sent out an email to my email contacts asking them to connect with me on LinkedIn. When too many of my contacts complained about that email, LinkedIn said I was not abiding by their rules. Fortunately, when I begged them to re-instate my account, they did. Never again did I opt to have LinkedIn connect with my email contacts!

Facebook changes its terms all the time. They've barred cli-ents for several days who've sent one too many messages, or friended a few too many people. They've even asked for pay-ment to message other Facebook users. You may think you're abiding by the rules and the next day they're shutting you down for messaging too many people. Also, Facebook has the ability to delete content, "likes," followers, etc.

Consider yourself forewarned about the dangers of relying too heavily on social media as your main platform. Instead, I suggest you think of social networks as your local gathering spots. If you've built your strategy out to include having a rela-tionship with your tribe via email, group programs, and your own website or blog, you'll thrive regardless if the gathering spot (a.k.a., social network) disappears, changes your access, or hides your content from followers.

You still have everyone's contact information, and communi-cation systems outside of the social networks. Plus you have your own website where your tribe can find you and you can share your latest content, schedule, and offerings.

Another way to think about it is to design your social media strategy to drive your most important connections closer to you. This includes attracting them as members of your email list, regular visitors to your website, as clients, blog subscribers, those who comment, contest participants, collaborators, affiliates, joint ven-ture partners, and overall online, offline, and in-person raving fans.

"Build your platform on your own personal website where you own the domain, you own the content, you have the control over it, and then use the social sites to drive traffic to the area where you have control," advises Neal Schaffer.

How To Bust Through

Here's the good news if you've already started growing your connections on the social networks: This audience is now *primed* for you to bring them "home" to your website, add them to your email and blog lists, and introduce to them other opportunities and offerings on your site and in-person.

Be sure also to inform your existing email lead list, current client list, and friends about your new social media presence. This way you have a web of communications and in the center is your website.

Coach Laura's Breakthrough Questions

- What messages do you want to share with your social networks to bring them to your list?
- How frequently will you email them?
- What value will your email messages provide to your list?
- What do you need to change or update on your website to make sure your site is the communication hub for your business and news?

Myth: Social Media Is A Wasteland Of People Sharing Superficial Content (i.e., Pictures Of Their Cats)

One of the biggest myths out there—as Joel Comm states—is that social media "is a wasteland of people sharing pictures of cats and their food."

Yes, admittedly, this is present and if you look for it, you'll find it: cats, dogs, gossip, and the lighter side of life. But that's not all there is to social media!

"Initially, I heard 'oh, Twitter and Facebook are places to show friends that your kids are teething,'" shares Ted Rubin, "and that Twitter is a place people mention when they're going to the bathroom." Though he originally bought into the superficiality of social media, he quickly got over that one.

Busted

The bigger question is what will you share? The best way to combat the "superficial content myth" is NOT to be superficial online! If you follow people who post superficial content on social networks, then you'll "see" a lot of that content.

Even though plenty of people share what they had for lunch, cute cat photos, and who they hate most, it will benefit you to realize that to position yourself as an influencer and "go-to" authority online, you must behave in alignment with that thought.

If the purpose for your being on social media is to engage with influencers and those who have interests in, and/or solutions about, your field of expertise, your passions, or current related trends, you probably won't engage with the superficial posters, as these people are not your audience (and visa versa.)

Many people use Facebook primarily to keep up with family and friends. These are also the people that find their favorite brands and "like" those pages. Teens and young adults use Facebook to stay connected with friends, and to find out about cool events and trends.

The good news is that the people you want most to engage with are connected to similar people, which means they are easy to connect to as well.

"Some people think they have to be on the same page as the people they want to connect with. In other words, if someone has a huge amount of influence based up on a Klout score (online influence rating) or reach, he/she may think 'maybe I can't reach out or can't add value. That's a mistake," says Larry Benet, Founder of Speakers and Authors Networking Group (SANG).

How To Bust Through

There are all kinds of people on social media. Your job is to connect with your *tribe* and the people you want in your tribe in meaningful ways. To do this, think about where you'd show up in the real world to meet influencers, prospects, and colleagues.

Make a list of leaders in your industry you want to get to know better. Start connecting with them online.

"I think anyone can bust through to connect with anyone as long as they take a little time to do plain old fashioned research. Then it may take a little creativity and persistence and it may take an introduction or a person they know in common, but if you do your homework and do good research, I believe that if you stay in touch, reach out and connect in meaningful ways all do to the power of social media. You see it all the time, when a military person invites a celebrity to a charity event and she actually shows up.

"There is so much insight and relevance in what people say on their social media. If someone is really trying to connect and not sell products and services (even if ultimately that's their goal), social media plays a role in giving real time intelligence in what's going on in someone else's world. You can see what's important to them. Sometimes they post it on Facebook, sometimes on Twitter, other times LinkedIn or Instagram or various social media,": Benet advises.

For example, Kathryn Rose, bestselling author of 9 social media books and sales professional, uses this social selling strategy to connect with prospects all the time. "There was a client that my colleagues could not get a meeting with. I found him in one of my LinkedIn groups where he had asked a question. So I answered the question and immediately reached out with a connection request on LinkedIn. He then accepted my connection request. I started searching for any kind of synergy and connections in common. Turns out I knew some people at a company where he had worked. He did end up also joining Twitter. He mentioned he was going to a conference that I was also going to be. So I messaged him on Twitter to let him know I was going. We ended up meeting and he became a client."

"It's not just about sending out a connection request, you need to meet people where they are in the buying journey. Often times you can see where they are at by what questions and content they are posting," adds Rose.

You may also want to connect with influencers' most avid followers and learn about them. Then look for groups to engage in on LinkedIn, Google Plus, or Facebook. As you start connecting with people and communicating, you'll become more visible and people will show up to connect with you.

The more people you connect with on social media, the more systems you'll need to manage your relationships. First, you'll want to filter the noise from relevant conversations. Facebook, Twitter, and Google Plus all have list building features that can be used as filtering tools to follow particular people's posts. Specifically, you can create "friend lists" on Facebook and include only the people whose posts you want to see.

Twitter allows you to do the same thing, and Google Plus circles can be used similarly. Create lists/circles such as industry leaders, prospective clients, and market influencers. Then, when you spend time online, look only at posts from your filtered lists, rather than on your home feed. Look out for what is important to these people and how you can be relevant there.

This will make your engagement time VERY efficient. In addition, you'll see little or *no* superficial content and you'll create deeper more relevant relationships.

"I was watching a Tony Robbins video, How To Quit Stuttering in 7 Minutes. I had stayed in touch with a woman in the real estate industry for more than 15 years. I know her son who is in college still stutters. So I sent the link to the video and told her 'check this out, I was thinking of you and your son.' She watched the video and she was in tears. At the end of the day that was something that was really important to her, that is her son. That is one of the key secrets to connecting," Benet shares.

Coach Laura's Breakthrough Questions

- Who are the leaders in your industry you want to form relationships with?
- Who are other leaders that your market follows?
- On which networks will you connect with these leaders?
- Where will you join groups in which your market participates?

- What is important to these influencers personally or business-wise?
- What resources, information or action can you give them that will be important and meaningful to them?

Myth: There Is No Return On Investment (ROI)

The media focuses often on reports about the lack of ROI from social media. More than that, the rumor mill is filled with stories about how 'I tried social media and spent a lot of money but saw nothing in return.' Many businesses want to see their social media marketing investment in time and money translate directly into an increase in sales.

If they don't see a sales increase as they increase their social media activities, they come to the conclusion that there is no ROI in social media marketing. They may either give up, or decide not to get started—especially if they've heard dismal news from other small business owners.

Busted

According to the *2014 Social Media Marketing Industry Report* by Michael A. Stelzner of Social Media Examiner, 50% of all marketers surveyed saw an improvement in sales due to social media marketing.[1]

Additionally, this report reveals that, "a significant 92% of all marketers indicated that their social media efforts have generated more exposure for their businesses. Increasing traffic was the second major benefit, with 80% reporting positive results."

The bottom line about the social media ROI misconception is that social media is not a direct sales tool. Rather, it is a relationship-building platform. When you effectively build relationships,

---

[1] Stelzner, Mihchael A. "2014 Social Media Marketing Industry Report." Social Media Examiner. Last modified May 2014. www.socialmediaexam-iner.com social-media-marketing-industry-report-2014/. sixth annual social media study, more than 2800 marketers reveal where they focus their social media activities, which social tactics are most effective and how content plays a role into their social media marketing.

business happens. Thus, if you can build a loyal community of raving fans, you will profit because of your growth in visibility and connectivity on social media.

"I get a 100% of my income as a result of my activity on Facebook and Instagram." ~ Tar'Lese Rideaux

How To Bust Through

If you're not getting more traffic to your website or new leads from your free offers, one of two things is happening, according to Joel Comm:

"You're either not bringing value, or your offer is not appealing enough to people. When you're bringing value to the conversation, people like, know, and trust you. When you have something to offer that they like, people will buy it. It's organic in that way.

"If they're not, then you need to ask yourself 'am I selling too much, am I selling too hard, am I not bringing enough value? Is my offer not compelling enough? Do I need to tweak something on it?'

"There's no one answer," Comm explains. "I've had offers that have had terrible results. It's not that people don't want to hear from me. It's simply that the offer didn't meet their need.

"When we look into businesses that are successful, they're there for the long term. It's not like you do this for a week, and then expect to see this huge increase! You've got to develop the mindset that you're in this for the long term.

"As you know, relationships are not one night stands," says Comm. "They're about building a 'like, know, and trust' factor with people so that eventually a transaction takes place. That transaction might be the sale of a product or service, or it might be the exchange of a philosophical idea.

A sale is somebody adopting your way of believing. That can occur through an exchange of money, or through an exchange of ideas."

World renowned connector, Larry Benet emphasizes, "Relationships are important and so you need to invest into them. The best way to invest into them is by understanding what's important to someone else and then connecting to them online, email, Twitter, Facebook with resources that will ultimately serve them, their goals, hopes, dreams, desires, and aspirations.

"You could reach out to almost anyone and say, 'Hey I'm not trying to sell you anything, but I was curious I'd like to get to know you better. I have a fairly influential network. My goal this year is to serve more people. What is the most important to you right now? Either personally or business-wise in case either I or someone in my network can help you.' Just by being a bigger giver you will see results.

"There is a huge correlation between the value someone puts out in the world and the financial remuneration that comes back to them. That tends to be an equation that holds true. Think strongly about as you add value to more people."

Coach Laura's Breakthrough Questions

- What connections have you made that you could re-establish a conversation with?
- What value is your audience appreciating most getting from you?
- Are your offers and posts getting good response?
- How can you take 1 relationship each week deeper?

Myth: Social Media Marketing is Free

Social media companies let you set up accounts and post content for free. You can get a fair amount of traction from posting engaging content, as previously described. Also, they provide widgets for your website for free.

So, yes, it's free to use, which is a huge benefit and makes social media one of the most accessible marketing tools out there. But will holding on to the "I want business for free" expectation serve your business?

Busted

At minimum it's costing you in time or in the money you are paying someone to develop your social media marketing content. Many people approach me to ask if I can "do" their social media for them. However, they have a "this is a free platform, so it shouldn't cost much to use" paradigm, and typically shy away from wanting to invest in someone to help them out.

In effect, by refusing to invest personal time in, or a budget on, content creation and quality engagement on social media, they thwart the extent to which their social media marketing will thrive.

Other costs involved in implementing a marketing plan online make good sense as well. Since *all* social media networks are very heavily visually-oriented—ever since Pinterest skyrocketed in 2011—you can miss out on a huge branding opportunity if you don't have a professional graphic designer on your team or one who's readily available to you.

These days many virtual assistants are adept at creating really wonderful looking graphics. The advent of apps and online photo editing sites has made superimposing quotes over images super simple!

Canva.com, PicMonkey.com and Quozio.com, are a few good resources for creating graphics—and they're free to use. Once again, however, consider hiring the *right* people to help you create great looking images on a consistent basis. This means they're thinking through what makes valuable content to help brand and promote your business.

There's also social media advertising that does cost actual dollars. Facebook has made advertising a necessity for business page visibility. Many now say that Facebook has become a "pay to play" network.

"Facebook, of course, costs money—real actual dollars," points out social media consultant Andrea Vahl, and co-author of *Facebook Marketing All-In-One For Dummies*. "Facebook's per page reach has declined dramatically, so you're going to have to spend money to boost your posts each month. You're also going to have to look at spending some money to be a little bit more effective there," she advises.

"I like spending money on efforts such as growing your email list because I think that growing your "likes" on Facebook now isn't quite as attractive as it used to be," Vahl notes. "That's because your "likes" people are not going to see your posts often enough.

"Instead, focus on getting those same people onto your email list. Think of Facebook as a real database of potential clients that you can target specifically and really well with Facebook Ads."

Social media marketing is extremely cost effective when done well.

How to Bust Through

Set a budget of time and dollars. Be wise about this one. First, clarify your goal and then identify which networks have the highest payoff for you.

Perhaps you're like, business storyteller, Karen Dietz, who doesn't like to invest her time in writing blogs. However, she does love reading and finding great articles. Once she started curating content things began to take off for her.

"I use Scoop.it which is a platform for content curation. There are other platforms out there. Find the one that works for you. I find great material out there on the web, bring into a platform and do a review or microblog in ways that are useful," Dietz explains.

"Curation is very different from aggregation. Many people are really just aggregating content and calling it curation. Aggregation is the feed that collects the tweets and article links on a topic and puts it on a platform without comment, filter, or review. That is not nearly as useful or valuable as curation. We are swamped with information. How are we going to make sense of all that? For example, if you Google storytelling you'll get millions of results. That's not really helpful." Dietz positions herself to, 'let me help you sift through, find the best ones that are useful, and share highlights of how it may apply.' Then she sends that reviewed content out on all social networks. "It provides great value for people. They don't have to spend time searching if they follow me," says Dietz.

You may want to integrate curation to lessen the burden of content creation. As we go along in the book, notice that **the primary elements of a great social marketing strategy include content creation, curation, engagement and tribe building.** As you craft and implement a strategy track the results and then tweak the strategy. Then track your results again. Over time, this testing and tweaking will save you time and money.

Coach Laura's Breakthrough Questions

- What are your social media marketing time and dollar budgets?
- What areas would be of value to research on behalf of your market?
- Is your social media branding at the professional level and consistent with *all* your marketing materials?
- What are the *highest payoff* social media activities you are willing to pay someone to do?
- Do you have a social media advertising budget?

Myth: If I "Like," Friend, or Follow People In My Line of Work, They May Steal My Clients

Allowing only one person from each professional specialty to join a business networking organization chapter is a popular small business model. The theory is that, without competing professionals in the group, more referrals will go to the one individual in that specialty. Further, members won't be conflicted about who to refer.

In other words, if I'm a business coach and there's another coach in the group, fellow members will limit referrals they'd give to me because they want to split their referrals fairly.

If networking was a math game, that would be correct. However, business by referral is not that cut and dried.

Many people apply this same thinking to social media marketing. They seek to limit their connections out of fear that their competition may see their prospective client base. If they like the page of a colleague, or share a colleague's blog post, they're fearful their clients and prospects might hire the competition.

Busted

Fear is not reality! Thinking that if you don't engage with your competition you'll prevent clients and prospects from hiring them is totally fear-based. If I'm in an offline business networking group that limits members to one per specialty, members may still have other professionals outside of the group who they prefer to refer clients to—rather than *only* referring them to someone in the group.

The integrity of these kinds of groups falters in that way. However, imagine if there was a networking group where several coaches and I come across an opportunity to coach someone in a profession with which I have no experience. Most likely I'd refer this client to another coach within this group. Opportunities for the coach would actually increase!

Karen Dietz is a corporate storytelling consultant, avid Scoop.it content curator, and author of *Business Storytelling For Dummies*. She generously shares regularly, and reviews others' corporate storytelling advice, articles, and website links. Doing so has improved her business dramatically.

"I've met so many great people who are my competitors, and established relationships where we share and advocate for each other," Dietz recounts. "No one does my business like I do it.

"I'm not doing *their* business—I attract the clients who want to work with me, and there's no scarcity of clients. When I share my vision and what I do best, the rest follows. I have been invited to author books, lead webinars, and give a Tedx talk—and I feel that's just the tip of the iceberg! Who knows where this will lead? "

"There's no such thing as competition online," my social media mentor Mari Smith has been quoted as saying. "It's now called 'coopetition.'" Coopetition is a hybrid word made up of the words "cooperation" and "competition."

Collaborating with your competitors is a viable approach for business growth. When organizations work together, more is possible.

How To Bust Through

A willingness to seek out colleagues in your industry will more than likely result in the knowledge that no one is doing *exactly* what you do. The perfect clients for *you* are not necessarily the perfect clients for others. You may be able to form strategic alliances with professionals in the same business in order to cover each other when on vacation, or to create a joint product or service.

The key is to come from a point of view of *generosity*. "Give first" is the first rule of networking. Generosity implies giving without expecting anything in return. What happens when you're generous?

- You feel good, and you spread good will!
- You help, motivate, and inspire others!
- It always comes back to you!

When you're generous, you're likely to be more satisfied with giving and the resulting feelings you experience. If you're not, it can be a really good indication that other things are blocking your success. Once you can be generous without feeling resentful or negative, then you're in alignment with what social media truly has to offer. If you're afraid of missing out, or of being taken advantage of, your social media marketing will backfire. In short, you'll get what you're expecting.

Look for places to give. As you read online posts and comments think about what advice, acknowledgement, or encouragement you can offer. Perhaps what you read will inspire you to create a blog post or graphic of your own. Mentioning colleagues is a very generous act. Sharing their content with your audience actually makes you look good, and you become a curator of great content.

You create a win-win-win situation: your audience wins with valuable information; your colleague wins with visibility; and you win by growing positive feelings between yourself, your colleagues, and your audience because they now view you as someone *in the know*. When you do curate content be sure to add your comments about what you think is most valuable and/or include an additional point.

Trust that being generous is going to work out well for you! You actually have nothing to lose by being generous even if no one buys from you! That's because chances are that the goodwill you create will reap rewards that may be indirectly related. Many people call that *serendipity*.

Hence, using social media for good can lead to a life of serendipity. Is that ok with you?

Coach Laura's Breakthrough Questions

- Who are the top 25-50 leaders in your industry?
- How are they using social media?
- What are they sharing that is getting the most engagement
- How can connecting with them online bring you more visibility in your market?
- Where can you find valuable content about your topic of expertise?

SECTION 2

MYTHS THAT KEEP YOU STUCK

Myth: It Takes Too Much Time
(Another Thing I Don't Have Time To Do)

"Social media? I don't have time to put another thing on my plate!" This is a common response from rising entrepreneurs not yet into social media. I understand and respect it. I agree that entrepreneurs don't have time to waste on meaningless activities or distractions.

It's very easy to become distracted on social media—watching video after video, or finding new content to read. However, guard your time and use it wisely. In fact, learning how to leverage your time is a business imperative that, when taken seriously, can make your life a lot easier and your business grow a lot faster.

Think about how much a client pays you for your time. Your time on social media must be purposeful and beneficial for your business growth.

Busted

What exactly is it that entrepreneurs say they don't have time for? Can the small business owner afford to avoid spending time marketing? That's like saying "I don't have time for any more business!"

A social media presence builds credibility. Consider it a part of your public relations effort to show the world you exist, are a *real* company, and are accountable.

Making social media part of your marketing strategy is saying yes to being a viable business in today's socially networked world. If things eventually change and people no longer find advice, share information about their favorite people/products/events, make recommendations, give reviews, or click on ads from social networks, then you'll know it's time to give up on social media marketing!

However, in today's economy, if you're in a growth phase for your business, marketing in general must take priority. In fact, a good rule of thumb is that 60% of your time should be spent on marketing. If you've been focused on the development of a product or service that's terrific, but that alone won't bring in business.

Perhaps it's time to shift things around a bit! If you're in start-up mode, consider putting 80% of your efforts into marketing. After all no marketing equals no business. Social media marketing involves getting the word out about your business in a relationship-oriented way.

If you're too busy to build relationships with key market influencers and potential brand advocates, what are you actually saying about your business? It could be construed as: "I'm too busy to be successful."

That being said, marketing is more than simply social media. Consider *all* aspects of marketing and determine where and how you can best leverage your time. If your market influencers and potential customers are on social media, you need to be there too for credibility, accessibility, and profitability.

How Much Time Should You Spend On Social Media?

Let's hear from some social media masters on how they spend their time networking online:

"How many hours are there in a day, and how many could you spend on social media?" asks Joel Comm. "You need some sleep, okay, so let's say you sleep four hours a night—you could spend twenty hours online. Would you still have done all that it's possible to do in the realm of possibilities even if you're on social media all day? Even if you know they'll always be other sites and other conversations in which to engage?"

"Allow me to liberate those of you in the social media world, and remind you *it's ok to use social media however you want!* Don't let people "should" on you by telling you what you *should* and *shouldn't* do. And if they cry "you're not using Pinterest, wow!" and make it seem like Armageddon is coming, don't buy into that...

"Use social however you want to use it," Comm advises. "Social should fit your lifestyle, your lifestyle shouldn't have to fit social. It doesn't work that way. Yes, you can always do more, but you don't have to. You got to have a life!

"I prefer Facebook, followed by Twitter, " Comm adds. "I check in at LinkedIn to make new connections if there's somebody specifically with whom I want to connect. Every now and then I'll post a picture on Instagram, but rarely on Pinterest. I've got twenty-two thousand people who have me in circles on Google Plus, yet I rarely go there except for hangouts, because I don't want to and that's okay. So it's okay for you, too.

"Whoever you are out there, if you're feeling the burden of social media, the tyranny of the social space, then maybe walk away for a few days," advises Comm. "It's not going to go away without you, the Internet's not going to blow up or implode, and all of your friends are not going to unfriend you because you've been gone for a few days.

"Figure out how social media fits in with *your* lifestyle. *I only use social media when I either have something to say, or I have time to engage—that's it. Easy.*"

"Understanding how to be strategic about what you're doing on social media so that a little bit of time can create a much larger effect" is the way to go, according to Sierra Modro, dubbed by the media "Canada's #1 LinkedIn Expert. "That's something I had to really work through for myself—understanding what kind of time investment I was interested in making—and learning where my time is most effective.

For example, Facebook presented Modro—named one of the top 10 media bloggers by *Social Media Examiner—with* some challenges. "Understanding that vortex I can get into where suddenly I realize I've been on Facebook *only* for the past five and a half hours! That was maybe *not* the best and highest use of my time."

Ted Rubin admits to resistance to social networking—until he got on it *and got it.*

"My initial reaction was 'oh, another thing to do,' and now forget about it—now I'm on it twenty-four/seven! Like everybody else that's deeply involved in social now—although I got very heavily involved starting in two thousand and seven and two thousand and eight—I wish I started sooner.

"As soon as I jumped on it, I made a complete one hundred and eighty-degree turn. I said, 'oh my God this is me, this is what I've been looking for my entire career!' I wish this had been around when I was in my twenties. Oh what I could have done, because I've always been a networker. I've always been a community builder," Rubin enthuses.

"I like to say that *'networks are just a series of nodes, but communities are people that support and help each other.'* I'm the one amongst my friends who brings everyone together. My friends from every walk of life know each other—I want them to know each other. I spent hours of hard work making that happen when it was a lot harder, and I needed to call this one on the phone and then that one. There was no way to make even a conference call to get everyone together to plan something when I started doing this!

"So, for me this was 'oh my god I'm home, I've found my place!'" Rubin relates. "And then, as I got deeper into it, I thought, how can everyone *not* get this? Not that everybody will use it the same way I do or leverage it same way, but the opportunities are endless!

"When I hear a friend say 'oh no, I don't want people to know what I'm doing, I'm not going there,' I'm like ' first of all, what are you thinking? You don't have to let them know what you're doing, there's no invisible camera that photographs you!' *It's really up to you how engaging and interactive you want to become!"*

How To Bust Through

"It's like everyone out there who tells their kids, 'how do you know you won't like it until you try it?'" Rubin continues. "How many of us have said that over and over to our children?

"'Here, try this.'

"No mom, I don't like it!"

"You haven't even tried it!"

"I think the biggest myth is that people think things are not for them—but they need to go beyond that. Get in there try it!, he urges. "And here's the most important thing: *it takes time so don't expect immediate results!*"

Modro reminds the entrepreneur willing to give it a go to keep it simple:

"Honestly, it's easy to get caught up in the inertia of not doing it" she warns. "Even if you've made that decision, commit to it and set aside time—even if it's fifteen minutes. Even in that short amount of time, you can go find your audience, listen to your audience, and start contributing to your audience," she says, "and they will welcome you.

"If you prove yourself a trustworthy resource, the people with whom you want to work will want to work with you! But you have to create that environment. The first thing to do is get over the inertia of saying 'this is scary and I don't know how to do this.' Stop thinking 'I don't know how to do this,' and just get in and do it!

"You can't do it wrong," Modro assures her clients. "You can be ineffective, but you'll learn from your mistakes. People will be helpful if you believe they'll be helpful. People will be helpful and coach you along the way. Relationships that you can build are best started today.

"To be strategic on social media and manage your time, begin with knowing your audience and targeting your audience," Modro continues. "Your ideal clients or companies are undoubtedly on social media. Targeting those potential clients and customers directly means listening to them *first*. Your customers and r clients will tell you what their problems are and give you the opportunity to start solving them.

"Starting with identifying your audience and then listening to them creates that conversation even before you start contributing. Go into that relationship with a better understanding of what they need and how you can fill those needs. You'll then find the people with whom you most want to work, bring them into your tribe, and see how you can most effectively help them."

"If you're a CEO, you need to jump in and understand the medium, " Neal Schaffer strongly advises . "Read tweets and Facebook. What CEO out there would buy a commercial on TV without watching it?"

"It all starts with a plan," adds Gail Martin, social media expert and international speaker. "Because if you don't have a plan, you will muddle around out there and get sucked into other conversations.

If you plan to use ten minutes in the morning, ten minutes at lunch, and ten minutes at the end of the day, have an outline of specific focused actions for each network," advises Martin, the bestselling author of 14 fiction and non-fiction books. "Post on Facebook, reply to mentions on Twitter, and check messages on LinkedIn."

Ted Rubin suggests you "start to leverage your content by hiring someone to get it out there for you. If you have employees, how about leveraging all that incredible intelligence you have in your employees? There are people are out there creating their own brand and looking for help. You can help them by supplying them with content, and then help them to syndicate it."

"Get a social media manager," Cole advises. "You can get reasonable help. Attorneys and doctors have staff members who field appointments, and can do social media during their down time. Have them go to social media every hour on the hour for five minutes. Simply retweet to keep your business top of mind.

"This ensures that when a client needs your services, they think of you first because you're the business they saw last," adds Cole. "If you hire someone, start with two hours per day Monday through Friday. You can find college students, interns, moms who recently had a baby and want something to do on the side.

"You can get incredibly talented and qualified individuals to work with you for anywhere from ten to thirty dollars an hour in your time zone in America, someone who understands your culture," Cole concludes.

Julie Renee, a Brain Rejuvenation Expert who helps individuals to reach a balanced vibrant life, and author of *Your Divine Human Blueprint,* has a simple plan. She spends a total of two hours a week on social media.

"I go online three or four times a day when in my office. When traveling, I'm online twice a day. When I get overwhelmed, I go on only once a day," explains Renee, the 2010/11 National Association of Professional Women's Woman of the Year Award recipient, and 2012 winner of Powerful Women International's Global Leadership Award.

"My time on social media totals only a of couple hours a week," she reveals. "I've prepared one hundred quote image photos, so that when I'm in a rush I can still post beautiful images. Make your posts about personal things you are doing.

"People really love seeing photos of me with people, and I show them a little of my personal life." Recognized most recently by the Big Money Speakers community award, cancer survivor Renee mentors ambitious women leaders who refuse to play small but are being held back by illness or exhaustion.

As you can see, each small business can create its own customized social marketing plan. It doesn't have to take that much time.

Coach Laura's Breakthrough Questions

- Are you willing to observe what users of social networks in *your* target market are doing there?
- When in your calendar will you make time to do that?

- What do you need to get comfortable with listening to on social media?
- Who in your company will be spending the most time on social media?
- Are you willing to work with them to create a plan for using the social networks?

Myth: It's Not Relevant To Me And It's A Nuisance

I can't tell you how many people I've met who express such disdain for social networking, including for the largest network of them all. Many of these myths perpetuate because people look for an excuse not to get involved.

Remember, since we've already defined a "myth" as *a widely held but false belief or idea,* how can we get our minds around the fact that social media may not actually be a nuisance? If this *nuisance* can sky rocket your business success, why would you stay away simply because you feel *some* resistance?

This particular "myth" masks the truth. In other words, people have bought into a lie and are being left behind in the dust because of it. If some people love social media, find it stimulating, life-enriching, and an incredible form of communication and business development, how is it that this myth perpetuates itself?

What if *social media* isn't the nuisance you assumed it to be? Perhaps it's only considered so by the user who doesn't take the necessary time to understand and discover how it might enhance their life, their business, and the lives of others.

Am I saying social media is for *everyone* or every business? *No.*

Busted

Rather, I ask you to take pause. Get really honest with yourself and contemplate this feeling that social media is nothing but a nuisance. Is social media doing something to you to bug you? Or, are you annoyed because you're not making the most of it?

Perhaps it isn't relevant to you. However, do your clients and friends use it? Are your prospects there?

"As a business owner, you must stop taking Facebook—or any of the other networks—personally," Schaffer advises. "When you untie your own reactions and look at social media platforms as business tools, it's going to shift your mindset and help you succeed in social media business."

If you're not famous, and want to become known as leader in your field, it's time to embrace social media.

How To Bust Through

At this point, it's time to acknowledge whether your avoidance of social media is about your frustration because you're not getting results, *or* whether it's because you're not willing to put into place what it takes to use it as a business development tool. In other words, are you declaring social media an annoyance simply as a justification for feeling annoyed?

If you can see how clinging to this justification is costing you success online, you may want to give up the excuse so that you turn your *resistance into revenue.*

James Hickey, one of the top leaders in the internet marketing industry for business owners, entrepreneurs, and start-up companies, works with clients to "make someone *not* famous, famous."

"I use a combination of content creation, articles, videos, blogs, and press releases," explains the man who created the Internet Marketing Training Center to train internet marketing consultants in 2011 . "If you don't know anything about a person, how do you learn about them? You'll learn about them through content. Put out some YouTube videos about this person.

"Let's say you have an empty glass and I put in it a little bit of water and equate that to content online," Hickey elaborates. "It won't get much traction. However, if I keep pouring water eventually you'll get water everywhere. At first you don't get much traction. Once it's overflowing, then it's everywhere.

"And that person's content is on the web showing up in a YouTube search result, or a keyword search. It could show up on Facebook shares, Tumblr, Twitter, etc. Steady content helps you show up everywhere. It's like old school marketing where you see a brand doing all kinds of advertising for themselves over and over.

"Like Shell gasoline advertisements on TV, billboards, flyers, grocery stores," Hickey adds. "When you see a Shell station—versus Joe's gas—you pull into Shell because it's so familiar."

Coach Laura's Breakthrough Questions

- Do you feel social media is a nuisance?
- If so, are you really annoyed because you're not capitalizing on it?
- Is social media relevant to your prospects, clients, and colleagues, and do they participate on social media?
- If so, are you willing to give up your *reluctance for revenue?*
- Would positioning yourself as an industry expert or leader help or hurt your business?

Myth: I Want My Keep My Personal And Professional Information Separate On Social Media

If you've been using social media to keep in touch with family, this myth may be your biggest stumbling block. Perhaps you never even mention your work or share business-related content. Transitioning from using social media for a personal use to business use can be disconcerting. You may be thinking one of several things:

- I don't want clients and prospects to see photos of my family
- I don't want clients and prospects to see what my family members post
- I want to keep my personal posts separate from my business

This concern primarily comes up with Facebook, as Facebook remains the most personally-oriented social network.

Busted

Good news! For the most part, you can keep your personal content separate—or out of sight of all your Facebook friends. Your business-oriented content can be posted for the public to see. The key lays with the publisher, privacy settings, and actual content you post.

Create A "Friends List" For Family and Close Friends
If you create a "friends list"—including specific family and friends who you want to view your personal content—they will be the *only* ones who see it if you simply select this list as the *audience setting* when posting.

It may seem like a lot of work to create this list and remember to select the audience, but it's one way to keep your content separate. Instead of worrying about who sees what, consider making the majority of your content for public consumption.

Remember that if you comment or "like" any content that your family and friends create for a public audience, it will then be viewable by your Facebook friends. That activity is also available to show up in the newsfeed of your friends, and in your public activity. Also, if you're *tagged* in a photo one of your family or friends post as public, it can be seen by everyone as well.

Create A Private Facebook Group For Your Family And Close Friends

If you really want to button up the privacy with your family and still engage on Facebook, create a private Facebook group and encourage them to participate. When you comment on content in this group, only members of the group will be able to see it.

Regarding the visibility to others of what your family members post, they will not see what's posted except in these two situations:

1. They are friends with your family members
2. You "like," comment, or share your family members' content

Then your activity and link to their post shows up in the right hand ticker in real time—and possibly in the newsfeed—as you are doing it. So, if you keep your interactions to a minimum you can see what your family is posting and not have personal content exposed.

Each person must decide what's important about each social network. Evaluate your family interactions, and consider the possibility of expanding the reach of your work and of gaining valuable connections. I believe you can take the best of both worlds and integrate them in a way that works best for you.

Facebook business pages are the ultimate in separating personal from business. You can simply do all your connecting there. If you do not "friend" people from your personal profile, you can use advertising to connect with your specific audiences and leave your profile for personal use only. This will take a budget, but it's a great way to build your business page.

On Twitter and Pinterest you can get around this conundrum by having a specific Twitter account for your business and another one for your personal posting. Two accounts may be extra work to maintain, so consider if you'd rather have one account that you humanize with some personal content.

LinkedIn is primarily for business use, so if you really don't want to use it you can choose to use different networks for different purposes. If your market is not engaging on Facebook, then by all means don't spend time using your personal profile to make connections. However, a good majority of businesses have an audience that participates on Facebook.

You are the *brand ambassador* for your business. I call this role "Brandividual." Think of it like this: You are Miss America, and no matter where you go in public, you represent that role, you *are* your brand. If you're living in alignment with your values, and these are in alignment with your business, this is a simple task. If not, then seriously contemplate aligning them. It will bring you greater peace of mind and prosperity.

Also think about your employees as Brandividuals. When they speak about your company in person or it mention online (even in a LinkedIn profile), they are a reflection of your brand. Be sure to have a social media policy in place, so they are aware of the standards you expect them to keep, and how they can help support and enhance the reputation of your company.

Coach Laura's Breakthrough Questions

- What's important to you about spending time on social media? List each network and answer for each...
- Is it important to make connections for your business on Facebook?
- Are you willing to make friends strategically on Facebook in the hopes of developing connections that might lead to business?
- Are you willing to be a public brand ambassador of your own company?

Myth: There Are So Many Networks—It's Too Hard To Keep Up With Them All!

Yes, there are a lot of networks out there! It *is* tempting to think you have to jump on every trend and master each network. However, this mentality leads to overwhelm.

If you try to be everywhere, chances are you won't do well with any network. It's better to develop a social media strategy that addresses your most important networks.

Busted

No, you don't have to be everywhere. You can be only on certain platforms and, once you're set up there, do relatively little.

When a new network pops up, typically there's a rush by a certain demographic to that network. Just because your friends are—or aren't—there doesn't mean you should or shouldn't be there as well.

"You need to analyze each one as a tool," Schaffer reminds us.

Many social media managers in the first couple of years of Google Plus disregarded it because they didn't see their colleagues and friends adopting it, Schaffer reports. On the other hand, there was a rush to join the Foursquare network.

Actually, the savvy social media plan for small business will be one that reflects your understanding of the network, who's there, if they are your market, and how and for what you would use it.

How To Bust Through

Budgeting online marketing time is just as important as allocating dollars for *every* small business. Furthermore, a business's social media time investment needs to be purposeful and well thought out.

"What you really want to do is focus on just one or two places," recommends social media strategist Vahl. "People think that because they *did* Facebook now they have to *do* Pinterest *and* Google Plus *and* LinkedIn *and* all the other stuff, and that's where you can quickly get overwhelmed."

Preventing overwhelm is important for business owners and team members. If overwhelm sets in, positive results are difficult to produce and resistance grows. In fact, if you become seriously overwhelmed, it's best to cut back your activity on social media to one network.

"The truth is you have to do just one or two really well," explains Vahl. "And as you get that organized into your daily life, you can maybe add a little bit later."

"Just chill out—just be you," Vahl advises. "What do you want to talk about? What do you find interesting? What questions do you want to ask? What questions can you answer? Who do you want to engage with? What brands do you like? Just be you..."

"Because we're separated by miles and miles and miles of fiber-optic cable, and because we can connect with people on the other side of the world, often we forget we're surrounded by people," adds Comm. "We forget we can glance around and say, 'look there's a person right there!'"

Tina Dietz, international small business and lifestyle design expert, recommends "Pick the one you're most comfortable with and then just make it juicy for yourself. Make it fun and you can always add on from there.

"You don't have to be everyone, everywhere, every time to all people," Dietz agrees. "You just can't do it and it's not authentic anyway. A lot of it has to do with finding out where your ideal colleagues, ideal customers, clients are coming from and for a business to consumer business, Facebook might be a better fit for somebody as opposed to somebody who really does business to business work and might want to focus more on LinkedIn.

"There are going to be things that will point to the most natural fit. If you do a lot of work and a lot of business with generation Z or even Generation Y, you might want to get involved a bit more on Instagram and things like that. Chances are if those are your ideal people then you probably going to feel fairly comfortable being there yourself "

Let's face it—the small business owner has a lot on his/her plate. With a simple social media strategy, keeping up with it is EASY! Less is more.

Coach Laura's Breakthrough Questions

- What network will you start with first?
- On what networks do you want to maintain a presence but not actively engage?
- Which networks will you put on the back burner?

Myth: Social Media Will Create Too Much Personal Life Exposure or Damage My Business Reputation

Many people, consumed by the thought of being overly exposed on social media, shy away from its use because they worry that something bad will happen as a result of their participation. (Note: This myth is not relevant for the person who's already comfortable sharing on social media.)

"My clients' number one fear of social media is a public relations crisis," says Samantha Hartley, president and founder of *Enlightened Marketing,* and a former international marketer for *The Coca-Cola Company* in Russia and the U.S.

"Afraid of negative comments, they say 'we can't be on Twitter because we saw some insane flame war get started—if something like that happened, we'd totally loose it!'

"So they have an attitude of extreme danger toward social media," Hartley continues. "Their mindset is that they can't possibly share anything without risking damage to their reputation. Typically, these are businesses such as a printer or a coffee shop—not a hospital that has to worry about HIPPA laws compliance! Small businesses often display disproportionate fear. So much of what I do is calm people down."

Recently, I met a gal who expressed to me at a public networking event: "I just feel I don't want to put myself out there and be so exposed." This intense apprehension came about from her belief that if she started using social media, something about her that she didn't want out there would be revealed. A dynamic woman with a wonderful business, she nevertheless seemed quite comfortable in an in-person social networking environment.

Busted

I asked this woman how networking in person is different from networking online.

"It's just different," she replied.

"You only share here what you want to share with people, right?" I pressed her. "You have the same kind of control online—you only share what you want people to know, and nothing more."

"The world is only going to see what you want them to see." ~ Emmanuel Dagher

My goal was to plant within this gal a seed. I wanted her to look at social media more as a known entity than an unknown one in the hope that she'd feel less vulnerable. Could it be that she feels obligated to share more personal information than she wants to on social media? Or, perhaps she fears someone else will share something about her she doesn't want them to share?

Well, we run that risk in real life, too. Chances are that won't happen to the average business professional—even on their personal networks—if they're wise and live their life as if they have nothing to hide.

Next I gestured as if handing a steering wheel to my perplexed networking friend and said "you're in control of what you share about *you*, your business, and others. You are in control of how exposed you are. Now that you're in control, take the reins and go for it!"

If you relate to this fear of being so exposed on social media—or fear that you'll lose control by being on social media—I have good news for you: You can have the steering wheel, too! If you still feel uncomfortable about "exposure," it may be more about insecurity about what you should or shouldn't post.

Perhaps you think you might share something that will get you into trouble, cause someone to reject you or say something negative about you. If you're concerned that someone will act inappropriately toward you, rest assured that you can report them and block them.

"All of your worst case scenarios just inhibit your ability to connect with people in a rather convenient way."
~ Samantha Hartley

My rule of thumb is: don't post anything you don't want the whole world to see. The best way to ensure your privacy and safety, and still participate online, is to keep your private information offline. If you're on social media for business purposes, I highly recommend you skip sharing content that has to do with gossip, politics, or religion—unless you're in a business directly related to any of those areas.

Many entrepreneurs savvy about business realize also that they can't be promotional on social media if they want to gain visibility, popularity, and develop a loyal following. Instead, offering a human element to their social content is far more effective.

Often it's at this point they look like a deer caught in headlights and become confused about what to share. If this is where you're at, allow me to guide you...

How To Bust Through

There are several guidelines to follow when sharing. As a person on social media you'll want to offer some personal content. However, be wise and strategic about it. Share only what you want people to know about you—and that doesn't have to be very much.

You might share a little bit about your hobbies, or places you've visited. For example, my husband's hobby is glass blowing. We love glass art, and I'm excited to share his new creations every so often. People enjoy looking at this artwork and commenting.

Also, I love healthy gourmet cooking, so every once in a while I'll post recipes or something foodie-related. If I find an inspirational quote posted in an office or store, I'll snap a photo and share that. You can see this isn't private information. Yet it *is* personal, and allows people a little insight into me as a person. You can decide to do the same.

The easiest way to start posting is to share other people's content that's already positively received. You may also want to share links to articles you've read online that you find valuable, interesting, or heart-warming.

Retweet others' tweets that are in alignment with both you and your market. Share images and videos you find that move you on Google Plus, Pinterest, or Facebook. You may want to include a comment about your *take* on it. Can you see how that can make you feel at ease about sharing and maintaining a great social media presence without feeling exposed?

If you're going to post more personal content, be smart. Post content about places you visit after you are back at home. Why? Well, suffice it to say, you don't need to let the public know when you're away from your home! Keep your private information off social media.

If you're taking pictures of other people and thinking about posting them to social media, be sure to ask for their permission before posting. Never assume everyone in a photo is ok with your sharing it on social media.

Don't overshare. Many people ask "how often should I share?" The answer is a subjective one, and depends on the network and following you have. In addition, it depends on the quality of your content. Don't share just to share. Share with a purpose. If you have interesting and valuable information, or a thought or question that comes to mind, then by all means share it.

Minimum sharing would be once a day to each network on which you want to have an active presence. You can post more often on Twitter then on Facebook. However, since Facebook now shows only popular content, you can post more, but it won't necessarily be seen. So post your most engaging content and you'll be seen more often there.

Coach Laura's Breakthrough Questions

- Are you willing to participate in social media in ways that will positively impact your reputation?
- What topics do your clients/prospects enjoy reading about?
- How do you want to make a difference in people's lives or businesses?
- Given your clients' interests and your mission, list the ways in which you can bring value to your following online
- What blogs, news sites, or organizations do you enjoy reading?
- What hobbies do you enjoy?
- Whose content on social media do you enjoy?

Myth: Social Media Is For Young People Only— I'm Too Old And Not Techy!

MySpace is one of the early social networks that definitely attracted a younger crowd because of its music orientation. Facebook got its start as a college-oriented social network. In 2005, all that changed when it opened its virtual doors to the public at large.

Around that time, LinkedIn invitations started flying out to professionals, but since there was no widespread precedent, some accepted LinkedIn invites while others did not. Those who did accept LinkedIn invites often didn't do anything with them. Those engaged in social media early on were primarily younger people.

With technology getting into the hands of kids at the youngest of ages, there will always be a place for them online. Does that mean generations prior should shy away?

Busted

According to Pew Research Center's Internet & American Life Project in 2013, 72% of online adults reported using at least one social networking site. The strongest growth group was people age 50 and older.[2]

Social Media may seem like a Millennial's (people born in the 1980's & 1990's) playground, but Schaffer comments: "Older people who are very well-versed and experienced in business are typically crossing the social media chasm faster, and they're

---

[2] Watershed Publishing. "Social Networking Adoption Trends, by Age Group." MarketingCharts www.marketingcharts.com/online/social-networking-adoption-trends-by-age-group-35692/. Summarizing study from the Pew Research Center's Internet & American Life Project.

more effective at social media marketing. It's the 'new tools, old rules' theory in play. You're never too late to the party!"

As Facebook blew away MySpace and grew beyond college kids—and added visual enhancements—social networks became photo and video sharing spaces where everyone wanted to be included.

Grandparents started getting online to see photos of their far away grandkids, and to keep up with what their families abroad and cross country were up to. YouTube shot up as the number two search engine. Today it's hard to find anyone who hasn't ever watched a YouTube video.

As the masses poured in, so did the marketers. As long as the marketers focused on building relationships through these communication systems, they saw results. Professionals began to realize they could now stay in closer contact with their colleagues, reacquaint themselves with alumni, and get introduced to decision makers in organizations they otherwise wouldn't have been able to reach.

Authors, consultants, and educators now had audiences craving their wisdom training. Entrepreneurs started following their favorite business mentors. Personal development-minded folks found social media the place to stay connected to their favorite gurus.

Don't believe me? Look up the Dali Lama on Facebook and see how many likes, comments and shares he gets on every post! By the way, even Betty White is on Facebook and has a huge following. How old are *her* fans? All ages!

How To Bust Through

The best way to stay young is to stay open and up-to-date about what's new, and to do your best to let go of resistance. Resistance only keeps you stuck. When I lead a class, I tell people to relax as if their having a glass of wine, cup of tea, or a favorite beverage— anything that brings on a similar state of mind to watching a movie.

When content is presented that isn't clear, don't worry about it for now. Get excited and start implementing what you *do* understand. As you're ready, ask questions, look stuff up in Google, and come back to the more advanced concepts later. That's the beauty of replays!

"If you can click a mouse, you can do social media." ~ Laura Rubinstein

Digital media and expert coach Deb Cole admits: "I'm not techy. I'm not a young person who grew up with iPhones and swipe technology. In fact, I remember corded phones!

"I've found that whenever I've resisted something or been nervous about adopting things like blogging, or anything technical, it meant that my competitors were too," Cole reveals. "I made a commitment to myself early on that anytime I resisted, my coach had to push me and remind me to blast through it. Within thirty days I'd adopt it. And once I adapted to what's new and cutting edge, life was much lighter and simpler."

So the good news about resistance is that if you have it, so do others in your industry. And if you break through it sooner rather than later, you'll be light years ahead of your competition when it comes to benefiting from the use of new technology.

"Age is an issue of mind over matter. If you don't mind, it doesn't matter." ~ Mark Twain

Coach Laura's Breakthrough Questions

- Are people my age on social media?
- Are people my age my market?
- Where are people of my market's age participating on social media?

Myth: You Have To Be An Extrovert To Be Successful On Social Media

Does the word "social" imply outgoing and extroverted to you? If you're an introvert, you may have shied away from social media because of the notion that only extroverts value, enjoy, and are comfortable utilizing social media.

Busted

I believe that social media is better for introverts than for extraverts. Introverts typically process internally. On social media, you have time to think through your response, and/or exactly what you want to share, before actually responding. If you're a true introvert, you are energized by being alone, and drained when with other people. Good news: the best time to use social media is when you're alone!

Online Business Strategist and Founder of the Blog Squad, Denise Wakeman, is a self-admitted introvert who has worked online and used the web as a marketing tool since 1996. Wakeman explains, "as a very strong introvert I can control my energy and how much I'm participating online and what content I give out without giving away my privacy. I choose to post what I want and share what I want. There's plenty that people don't know about me. Most people, because I have been active for so long, feel like they know me very well. I choose very specifically where I want to reveal things and what I want to reveal. Social media is a very empowering place to be especially for introverts who may not feel comfortable in a live one-on-one situation which is a huge energy drain."

Furthermore, "there are other people who are introverts who are going to resonate with you, and if you can be really authentic and genuine about who truly are, your fans, customers, or clients will find you," points out Lisa Steadman, a relationship expert, author of several bestsellers, and publisher of the popular blog *Confessions Of An Imposter Mom*.

"They may buy your book, or your consulting services," Steadman says. "But it's really about being *you* and celebrating your natural style."

How To Bust Through

Perhaps you're a private person and don't want to reveal much about your personal life. That's fine! I totally advocate keeping your private information off of social media. Plus, *you* get to say what's private information. To keep going, ask yourself: what would make sense to share?

If nothing comes to mind, let me help. First, what catches your attention on social media? What really resonates or inspires you? Those pieces of content are great places to start. Simply share them. Most networks have a way to retweet, share, pin, and give credit to whomever or wherever you found it or, better yet, link directly to its source. Or, you can simply create an original post.

Also, be forewarned there are people out there who say unkind things, and that may happen. Lisa Steadman shares a really powerful example of how she dealt with a hurtful comment:

"I posted a photo of myself and a colleague announcing my excitement about bringing her back to my annual live event. In response I received a hateful comment from a man I didn't know: 'Wow Lisa, now I see why most of your photos are from the waist up...'

"Someone called me out on body issues about which I was already sensitive—my worst visibility fear was realized! I felt the pain—it hit me really hard and knocked me for a loop. However,

I realized I had a choice. He could be right, and I could shut down. I could delete his comment and pretend it never happened. I could respond with same venom.

"None of those responses felt right.

"Instead, I replied as I would to any post.

"'Wow I've never met you before, but I know exactly who you are, and I thank you for showing me who you are.' Then I took a screenshot of this comment and shared it again to Facebook with the following question:

"'For anyone who has visibility fears, what would you do if you saw this comment? Would you show up and step up, or would you shrink down?'

"Two hundred comments later, people shared their stories. Up in arms, they rallied around me," Steadman relates. "It was a beautiful illustration of how to stay strong and in my power. He didn't get to be the bully, and other women got to see that it's ok to be brave, bold, and not a shutdown woman.

"It's not ok to allow others to tell you who you are or what to share! You may get haters. Be willing to move through those difficulties. Know it's ok, and sometimes good to take a break."

Steadman became stronger and more influential as a result of facing head on this negative comment.

An easy way to start is with an intention to offer a positive contribution. The next thing to decide is if you're willing to become a public figure. Yes, even introverts can be public figures! This doesn't mean you give up privacy or become an instant celebrity—it means you're willing to stand for something and offer your say about it in any way that feels authentic to you.

Often it's introverted public figures who have the best reputations. They're not out there simply for public attention. In fact, they'd rather not be seen, but their cause, or work, and who they impact is important enough for them to make the effort.

Once you're willing to become a public figure, further your cause, share your talent, contribute useful information, or simply inspire others using social media, it becomes a welcome avenue. You may start sharing posts that you come across in public, or about events that are making an impact.

"Last fall I went from being simply 'a person' on Facebook to a celebrity," says Renee. "I took some advice from my dear friend Barbara Niven, a media trainer and Hollywood actress. She suggested using a behind-the-scenes strategy to up-level my game. So I took it on. I was going to meet the media at the National Publicity Summit in N.Y.

"A couple of weeks before, I started posting exciting things about who I was going to meet and taking pictures as I'm preparing for the trip, and I started to create a buzz. As I hit on the road, I posted updates from airport to airport. When I met producers of the *Today Show*, I shared a couple of sentences about what they told me.

"People received updates the whole time I was there," Renee reports. "That totally put me into celebrity realm. I got eighty to ninety comments on posts. I took pictures in the taxi and on the airplane, and people got really excited because it was behind the scenes. It allowed them to take the journey with me. " Renee's grounded social media presence allowed her to build trust with her audience, and to garner a high percentage of loyal fans and followers.

"I don't approach social media like it's going to make me money," she explains. "I approach it like I'm creating a celebrity buzz, excitement, and presence where I'm owning my bigness. I'm not trying to sell somebody something on social media— they know who I am. The website is on my banner. They know what I'm up to. I'm actually engaging them as friends and followers and people who are in my world. It's super fun!"

When you share the parts of your life and work that may enhance someone's day, you're on your way to breaking down personal barriers and growing your social media success. Renee went from someone who was afraid to go to a website for fear of being exposed to a computer virus, to using social media daily in order to make a difference in people's health and well-being. If she can do it, anyone can...

Client attraction expert Hartley shares that she's an extremely private person. "The language of social media is not intuitive for me" she notes. "It's like a second language. So it's a bit hard for me to share personal information. I don't have the need for everybody to know stuff about me. I've noticed, however, how much closer it's brought me to people when I do share.

"People do care about what happens to you—the thoughts you have and what interests you," Hartley add." Personally, I like to joke around, and make witty observations. Friends in my community enjoy that. That's what I try to bring to social media. On my business pages, I make sure we're in alignment with our brand. We tell inspiring and engaging stories, and encourage our community to share their stories, too."

Coach Laura's Breakthrough Questions

- What are your preferred ways to communicate (writing, audio, video, etc.)?
- What kind of impact would you like to have on your social media audience?
- What do your friends and clients ask you about?
- What hobbies or interests do you have to which others might relate?
- What social network posts do you find yourself drawn to?

SECTION 3

MYTHS THAT PUT YOU AT RISK

Myth: Images Found on Google Are Great To Post To My Social Media Accounts

Have you learned how to save an image from the internet to your computer? Wasn't it liberating to be able to do that? Then Google image search came along and you could find any kind of image you wanted. Some people love this search because they feel they can easily get images for their website. Right?

Wrong!

Imagine yourself posting an image you created (or paid a graphic designer to create) to Facebook, Twitter, Pinterest, or Instagram. Suddenly you see someone else posting it as if it's their own. They didn't give you any credit—they simply repurposed the image for their own use! Perhaps it was an innocent mistake because they thought the image—the one you created or had designed—was free for them to use.

Well, nothing could be farther from the truth!

Busted

If you take an image from someone else's website, or find one through Google image search—or even social media—and use it as your own, it's called *piracy*. Piracy is considered copyright infringement. All images—just like all written material—are automatically copyrighted to the creator when produced.

If this is news to you, and you have images on your websites, e-books, or social media that you have not officially acquired the rights to use, consider removing them from these sources.

"There is a difference between plagiarism and copyright infringement," explains attorney Denise Gosnell, Gosnell & Associates. "Plagiarism is taking something without giving the creator credit for it. But just because you give someone credit, in the world of copyrighting it still doesn't mean that it's not copyright infringement.

"The tricky part in social media is that if you're *taking* someone else's image—even if they didn't get a license for it—and that image was not free to use, but the person who created it didn't say 'I'm putting this out there for free for everybody to use,' then any time you 'save as' you're literally committing copyright infringement," Gosnell explains.

"There are exceptions to the copyright laws," she continues, "called fair use that we've heard about before. Now under the fair use doctrine you're allowed to put comments on it for criticism purposes, editorial, education, and classroom-type things.

The more it's along those lines, the more the courts will excuse it. The more you move it toward commercial use, and use it to make money or benefit your business in some way, the more they will hold it as copyright infringement without excuse. But at the end of the day, it's copyright infringement."

How To Bust Through

Gosnell continues, "I advise my clients to use *only* licensed images. Being more conservative in your social media strategy means buying licensed images of anything you'd like to share. Then you may superimpose your own quotes over it, and you have the right to create derivatives from it.

"That way you don't have to worry about someone shutting down your site down for copyright infringement. Copyright infringement won't necessarily get you sued, but it can result in your website being shut down through an image owners' filed complaint.

"If you're paying a contractor or an employee to create content, it's important to have two things in place:

"First, an agreement with your contractor that you *own* the work for which you are paying him or her. Assuming you want the right to make changes to it in the future, tell your contractor you want full ownership. You should consider this because, under copyright laws, the contractor owns the copyright unless he or she transfers it to you in writing.

"Secondly, make sure you include a clause in your contract that says the contractor will not use any copyrighted images without a proper license, or they will give you documentation showing that they have a proper license, or they will indemnify you if you get sued.

"But—and this is a *big* but—the problem is that even if they agree to that it doesn't mean they'll have enough money if you get sued because it was on *your* website and you're part of it. That is why it's really important not only that you put that in the contract, but also you have a way of policing what they're doing and have a submit process in place.

"You can then review and make sure that they are complying with the proper licenses, or you could get a subscription to iStockPhoto or Getty images and say 'I only want you to use my subscription. I can give back credits for you.' Now you can log in and verify that you have proper licenses for those images.

The tricky part is that the number of misconceptions out there. Just because an image is royalty free doesn't mean you don't have to pay for the licenses. *Royalty free* simply means that—once you have paid for the proper license—you can use it without having to pay additional royalties in the future." Denise Gosnell (more on the full video interview available at www.SocialMediaMythsBusted.com)

An additional resource for crafting your royalty free images is Canva.com. There you or your graphics team can create high quality images for free and low fees using their images.

Coach Laura's Breakthrough Questions

- Do you have a stock photo image account?
- Do you have a Work for Hire agreement in place with your graphic designers, web development team, and virtual assistants?
- Have you considered creating original infographics and quote images with your original images?

Myth: Social Media Is Great For Sharing Personal Information

Earlier we debunked the myth about avoiding social media for fear of being personally exposed. Those who share freely and abundantly, however, also need to consider the myth that sharing everything from *who you're with, what you're mad about, and when you're on vacation* is a good idea.

Because of the social nature of social media, it's tempting for some of us to post info about upcoming social outings, vacations, children's photos, people or companies who did you wrong, gossip, confidential industry intel, and private issues. Often we see people using social media as a platform to vent their frustrations. While this may offer some relief to them personally, think about the potential negative repercussions to their business and personal image.

Busted

If you want to grow your business, you'll want to be extremely conscious of what you post on social media. When you're on social media for business, you are there to communicate and connect with as many people as possible. If you have personal contacts, family, and friends with whom you want to engage, set up a private Facebook group and/or list that you share to because the majority of your content should be for public consumption.

There are already books being compiled with case studies about legal ramifications where content shared on social media has been used as evidence in court. That being said, it's fairly easy to avoid getting caught up in a situation that might damage your reputation or get you into a legal bind.

Here's a quick example Joel Comm shared during our interview:

"Recently I saw a post about a guy who was getting a divorce and he apologized publicly to his wife for what was going on. I messaged him privately, and said 'dude can I share my opinion?' I always ask people before I give my opinion...

"I told him 'I would pull that down if I were you. This is not for a public forum. This is a private thing between you and your wife, or between you and a counselor or trusted friends, but not to play out for the world.' Fortunately it did get pulled down, but I hate seeing that...don't do stupid stuff just because you can."

How To Bust Through

Simply go back to being clear about your message and your mission on social media. Remember the golden rule: do unto others as you would have done unto you. There are better ways to deal with an injustice from a company than to broadcast it on social media. Check with an attorney before you vent too loudly on social media. Some of these companies take defamation very seriously.

If you're working with a company (either as an employee, contractor, or client) be sure to ask to read their social media policy. You'll want to ensure you understand what pertains to you given your role as consultant, employee—even as a client. Also, if you are privy to confidential information, NEVER allude to or share it directly.

"Don't be stupid!" cautions Comm. "Seriously, don't be stupid about what you post on social media! Don't just post to post something. If you want to be controversial, go for it, be controversial. If you want to talk religion or politics, I don't have a problem with that, just be ready for the backlash when it happens.

"My general rule of thumb is don't post your personal relational stuff. I've watched people talk about their relationships with a boyfriend or girlfriend, and how this one did this, and that one did that. They get into it on social media. Everybody is watching it like a train wreck in slow motion."

This kind of content can also put you at risk personally because you may provoke someone to lash out or—worse yet—slap you with a lawsuit. If you keep your content original and positive, and approach social media to win friends, you'll be fine.

"There's a lot of personal information that's not really personal," says Modro. "The fact that I like shoes is personal information, but it's not going to be terribly useful for identity theft. There are lots of personal things you can share that make you more human and relatable. When you're thinking about what personal information to share, think about what's going to create connection," she suggests.

"As tempting as it is to complain, it's better to use social media as an extension of the best part of you," Emmanuel Dagher, intuitive and international bestselling author, advises. "If you complain, it's irresponsible, and what you're complaining about can really have an impact that may come back to harm you."

Coach Laura's Breakthrough Questions

- Think about a piece of content you want to share— what feeling or reaction do you have when you read it to yourself?
- Ask yourself, *why am I sharing this?*
- What overall feeling do you want people to have from reading your content?
- Does your content provide them with the feeling you *want* them to experience?
- If your content doesn't align with this, how can you change it to be more effective?

Myth: I Shouldn't Share All My Diverse Interests— If I Reveal Them All, I'll Confuse People

If you're alive, you have multiple interests, hobbies, and maybe even businesses! Do you feel stuck about knowing whether to share *all* of your interests or just your main business?

You're not alone! Many counselors, therapists, and professionals who deal with helping others are trained *not* to share their personal stories. Others, in even more conservative fields, may believe that their outside-the-box hobbies might not sit well with their social media followers.

Whether to share all your diverse interests or not is a very personal decision.

Is there a hard and fast rule? *No!*

Let's explore how to decide what's optimal for you to share on social media and how to do so.

Busted

Will you confuse people if you share your diverse interests? Not necessarily. You are human, with a life beyond what people know about you professionally. In fact, you may connect more meaningfully with your audience as a result of revealing your hobbies.

If these hobbies include things that people admire (sports, art, entertainment, do it yourself, travel, adventure, health, etc.), you're more likely to successfully connect with, than confuse, others.

Jenn August, Business Hypnotherapist and certified Success Coach talks about how—in her life and industry this issue comes up and how she deals with it. August shares "I was taught therapists are not supposed to talk about themselves. It's all about the client. Granted I'm a Hypnotherapist and success coach. None-theless, in my field, clients have to feel really safe to go into hypnosis and they have to really trust me."

When it came to social media, "I was able to talk about the business stuff, but it was very hard for me to post personal stuff that I thought would be of benefit. On top of that I'm also a singer, songwriter and artist. I didn't want to confuse them."

How To Bust Through

Perhaps you can relate to what August went through in the be-ginning. The key to breaking through is to remember that the social networks work best when you approach it as a two-way conversation. Being a singer, songwriter and artist, August was used to output.

"I started noticing the people's social media posts that opened my heart. I felt closer to them. I started noticing the things that were affecting me in a really positive way and made me feel more connected to someone I didn't know that well. I want to create accessibility and connection."

"Then I started eking out small bits of personal information and people really responded. They were really curious about what happened to my cat or how I came up with a certain way I was teaching. I became the person others wanted to connect with, and I learned how to dialogue. I noticed other people's so-cial media posts that opened my heart, and I wanted to create accessibility and connection."

"I created a focus to inspire people and invented *paper towel wisdom* where I draw little inspirational things on paper towels and share them. Turns out these are very popular. I also put mu-sic online. I really have stepped into realizing that people want to know who I am."

August coined the phrase *business transparency*. She advises, "People don't need to know everything. Instead consider what is in their highest good to know. What is going to inspire them? What is something I experienced that will benefit them? Now she feels free to bring out her silly nature, along with "what's in the highest good for them to know and will benefit them."

August invites us to take bold moves—"just a couple of molecules of something that is a little scary and you wouldn't have done the day before..."

Benefits of sharing your diverse interests:
This creates instant rapport and accessibility, which breaks down barriers and resistance toward you. They are more open to what you have to say and you become a natural filter for the right client.

"Inspire people and it comes back to you in multiple ways," August proclaims.

Coach Laura's Breakthrough Questions

- Think about your past week. What were the highs and lows?
- Is there something that helped move you through the lows?
- What did you learn?

SECTION 4

SPECIFIC SOCIAL NETWORK MYTHS

TWITTER MYTHS

M any people don't "get" Twitter. I didn't at first either. I find that people who don't get it are thinking too much. Twitter is one of the simplest social networks out there. This may actually be what throws some people off.

Think about it this way: Twitter is a public instant messaging system. It includes two ways to communicate directly with individual users. One is public mentions. The other is direct private messages. To speak publically to someone directly, simply use their @Username in the tweet. To private message someone, they must be following you.

That's all there is to understanding the concept. How this can work for you is where it gets exciting. With consumers, business professionals, the media, brands of all kinds, and business leaders all over Twitter, you now have access to communicating directly with your market and with the influencers in your market. Imagine the possibilities!

As I walk you through the following myths, you'll see how

you can become a well-known thought leader and go-to expert, and find new opportunities for business.

Myth: You Should Follow Back Everyone Who Follows You

Like most social networks, Twitter has evolved and made significant changes. Thus strategies for using Twitter have followed suit. Early on, following someone who followed you on Twitter was a kind gesture of acknowledgment. When someone follows you, which means—in theory—they are interested in your tweets and what you have to say.

Previously, Twitter allowed you to use tools that would automatically follow back the people who followed you. However, since mid-2013, Twitter disabled that function. Also, in order for someone who followed you to be able to direct message you privately, you would have to follow them back.

Keeping these things in mind, we can begin to see why it might not make sense to follow them back automatically. Also, because of the auto-follow tools and direct messaging feature, Twitter has become a haven for spamming.

"In the beginning, I believed that if people followed you, you are supposed to follow them back," says Comm. "So if seventy thousand people were following me, I'm going to follow seventy thousand back.

"Here's the problem: it became unusable for me as far as interacting and engaging, so I was one of the first ones that 'mass un-followed' everybody on Twitter.

"When I blogged about it, of course people began an uproar about 'how can the social media expert un-follow everybody?' But, if your stream is unusable then what's the point? So I un-followed everybody, and only followed the people I actually wanted to follow. Not that the other people weren't worth following. It's that I just didn't know who they were and what they were talking about or why I should care.

"So now I have about seventy-eight thousand people who follow me and I follow nine hundred back, and my stream is actually usable. I can go and see tweets from people that I know are not complete strangers simply filling my stream with their thoughts."

How To Bust Through

The good news is that there are several really good tools out there to help you use and manage your Twitter engagement. Following people simply puts their tweets into your Twitter stream. Personally, I don't rely upon the Twitter stream because I have a lot of people from the past who I followed but who are no longer posting relevant content. That means my Twitter stream is a bit noisy.

I could spend a lot of time un-following people to reduce the noise, but that seems like a waste of time. Instead, I use Twitter lists. I add key people to my lists and keep up with these lists using Hootsuite, a social media scheduling software.

In fact, Hootsuite is great for doing everything you can do inside Twitter, but in an easy to use fashion and on one page. You can retweet (and customize the retweet), schedule content to multiple Twitter accounts, and reply to tweets that come into your favorite lists—all without jumping to other pages.

"Using hashtags and keywords is the way to meet like-minded people to niche into your particular segment. For example, I'm a publishing consultant and I do editorial work. I used the discover symbol which is the hashtag (#) and followed that with the word publishing in the search field. I then saw whose names popped up. I would go to their feed and see if they were interesting and I followed them. If they were superstar to me, I put them into my list," according to Stephanie Gunning, Founder of Lincoln Square Publishing.

Regarding managing users, a tool like Manage Flitter works to un-follow people who are no longer relevant and who don't follow you back. You may want to continue to follow industry experts and thought leaders, despite the fact that they may not follow you back.

Myth: It Matters How Many Followers You Have

When you have a lot of Twitter followers, it looks good—it's like a social media status symbol. Initially, people may think you're really popular and have something to say. If you're looking for status *only,* however, then having a high number of followers is good enough.

But status doesn't mean business success.

Busted

Take a closer look at who your followers are, and the benefit that provides you. This is what's most important. I've had more than 30,000 followers for quite some time now. In the beginning, it was easy to get followers because a large percentage of people on Twitter followed you back.

"The number of followers doesn't matter," agrees Modro. "What matters is the quality of the connection you have with your followers. If you've got a hundred people who are all rabid fans of you, your products, and your services, that's what matters. Since they're on social media and they also have a life, they know people in real life and can talk about you offline in real life. Creating these relationships gives them a reason to talk offline about you."

"Real relationships happen on Twitter and relationships are what create business." ~ Sierra Modro

How To Bust Through

Give up the need to get a certain number of followers. Instead, commit to spending your time on Twitter to build quality relationships from the followers you already have.

Myth: Is Twitter Too Noisy to Get Any Business From It?

I have a feeling I might be just like you when it comes to Twitter. I open it up online or on my mobile phone, look at the main feed, and it's overwhelming. It's also distracting. Often, it's not even relevant. I wonder *how did these people's tweets even show up on my feed?*

That being said, my thinking may diverge from yours here: *I don't care what shows up on my feed!*

Busted

Just because it's noisy doesn't mean it's not useful. In fact, the noise factor means Twitter is more popular than ever as a communication tool.

Twitter consultant and social media expert Holly Kolman shares a powerful story about how she recently helped a company get to speak with the decision makers using Twitter. This company wanted work with business brokers. Their sales people had been cold calling but weren't getting their calls returned, or even acknowledged. In fact, they were hung up on regularly.

Kolman started a new Twitter account and began to follow the people they needed to reach. She worked on building rapport by following key business brokers there. She retweeted them, and mentioned them in her tweets. For about two weeks, she did this with a particular business broker who seemed active. At one point he replied and thanked her.

By the end of the second week, he asked what Kolman did. Notice how he invited the conversation. Kolman never directed him to a website, asked him to come to a landing page, or did any kind of selling. Their connection was based entirely upon relationship.

When Kolman asked for permission to call him on the phone, he agreed, and when she called, he was expecting her call.

"I know you, we're Twitter friends," he told her, and she was able then to explain her company's benefits to his company. He agreed to accept a call from the president of Kolman's company to talk about their product.

This approach flies in the face of what most people think about Twitter. When you use Twitter as a tool to develop and manage relationships, it helps strengthen your connections. If you're selling magazines, it would be a different strategy. But for higher level sales, building relationships is key.

It's important to be careful not to irritate people, to read their signals correctly, to discern whether they are willing to engage— or perhaps they're not the ones tweeting.

How To Bust Through

Recognize the value of Twitter as a means to get in touch with targeted people you otherwise wouldn't be able to reach.

"If you use Twitter as a relationship building tool, it's like calling out someone's name in a crowded room," explains Kolman. "The person who needs to hear it will hear it. That means you must do your research and find the people in your market—or the best people—who will refer you. It might be the media or bloggers—the more you build these personal relationships, the more you'll be able to cut through the noise."

Coach Laura's Breakthrough Questions

- Who are the decision makers with whom you want to speak directly?
- Who are the influencers connecting meaningfully with your targeted market?
- How can you be of value to them?

Twitter Action Guide

1. Identify keywords for finding people on Twitter
2. Do your research and follow the people you most want to reach
3. Engage them strategically by retweeting a couple of their posts weekly, and reply to their interesting tweets

FACEBOOK MYTHS

Facebook is the mother of all social networks. When people think about social media, Facebook is typically the network that comes to mind first. There is good reason for this. Facebook is currently the most popular social network out there, closing in on 1.5 billion users. More than 1 billion of these actively use Facebook on their mobile phones.

Earlier in this book we covered one of the biggest myths about keeping personal information separate on Facebook. Now that you've resolved that issue, you can begin to use this network effectively. The goal of this section on Facebook is to prevent you from making common mistakes and to get you to take action that builds your small business.

Myth: Accept Every Friend Request That Comes Your Way

Some entrepreneurs feel it's in their best interest to accept *every* friend request that comes in to their personal Facebook account. The myth here is that everyone who wants to friend you on Facebook is a good fit for you.

Certainly it's a good idea to grow friends, but does that mean you should accept ALL requests? There are many friending practices and philosophies out there. Let's review what really serves the small business owner when it comes to your personal Facebook account.

Busted

I recommend strategic friending, even though this is your personal account. Remember you are a whole person with a business life that's part of your personal life—which means you can use your personal Facebook account to make new friends who may turn into business alliances, referral sources, and maybe even clients.

After all, the nature of networking is to develop relationships first and allow business to happen second, as we've been discussing all along. Thus, friending people you really want to get to know because they have a common interest (relevant to your business products and/or services), or because they are in your client demographic, makes sense.

You may want to be a bit cautious if you see an onslaught of friend requests coming from people from a different country. First, click on their profile. Notice if they have many friends. Do they have a bio or contact information? Their posts should show tell-tale signs of whether they are an active Facebook participant, or simply building a fake account. Are they interacting, posting personal photos, or sharing information? If not, skip friending them.

"Facebook was never designed to accept every friend request." ~ Joel Comm

"I conducted a mass un-friend of my entire list, even though this offended some people " relates Comm. "I wrote a blog entry explaining why I did it. Then, I re-friended the people I actually know. Now I'm directly connected with just over one thousand people on Facebook.

"Of course I've got my fan page, which is for mass following…I've got over thirty-seven thousand people following my fan page. And I have it opened up so that the public can subscribe to my personal feed, onto which I post a great deal."

How To Bust Through

Above, I spoke about strategic friending. This concept begins with the intention to be available and open to serving people through your expertise. Once you have this intention in place and you let go of business expectations, things begin to open up and people don't see you as desperate. Rather, they view you as generous.

To start accepting friend requests strategically, connect with colleagues, alumni, friends from the past, and like-minded individuals. Build the relationship from there. Think about your "ideal Facebook friend." This should be someone who engages regularly on Facebook, something you can tell from their posts.

Compare your ideal client profile to the person requesting your Facebook friendship. Compare demographics (age, gender, occupation, and geographical location) to determine if there's enough synergy for you to accept their request. Remember, you can always un-friend someone.

Keep in mind that Facebook caps at 5,000 the number of friends you can have. However, they created a feature called "Follow" which you can activate so that anyone who wants to can follow your publically shared posts.

This eases tremendously the pressure of accepting friend requests. Now, when someone simply requests to Add You as a Friend, they are automatically following you. So if you never respond, they still may see your posts show up in their feed. Accept the requests from people you want to engage with personally. Those you want to keep track of you can simply follow.

In the beginning, you may not be getting many friend requests. This is where you want to start exploring Facebook to make friend requests of your own. The best way to do that, after you've connected with your colleagues and other industry leaders, is to take a look at who's following them and see who their friends are.

You may make friend requests of these people, but be careful not to fire off too many of these requests at one time. Facebook guards against spammers. They have been known to shut down accounts or prevent friending for a period of time if there is too much repetitive activity (like adding friends.) Alternate your activity by accepting a few friends, and then engaging them and your other friends.

Engaging your friends can include actions such as:

- Thanking them for connecting, wishing them happy birthday
- If you meet them in person and they are ok with you posting a photo of them, tag them with a "nice to meet" you mention
- Comment on their posts
- Share their posts that resonate with you

Myth: If I Get 10,000 "Likes" On My Business Page, I'll Get Sales

Having a lot of "likes" is another social media status symbol. It may be good to impress your colleagues, but it doesn't mean an increase in business. Anyone can pay for "likes"—you can do that by advertising, or by trying to game the system. I don't recommend the latter. Just having "likes" in and of themselves is not going to bring business in your door.

Busted

If you're a local business, or in a niche market, *a lot* of "likes" could mean a few hundred. You don't need 10,000 "likes" to create a thriving business. Furthermore, Facebook page visibility has gone down so dramatically that sometimes it may not even be worth it for the small business owner to do much more than make sure he or she has an established and maintained professional presence. That way, you're there, in case anyone proactively seeks you out or simply stumbles across you there.

In fact, with Facebook's follow feature on personal profiles, it is often acceptable simply to keep a personal profile as your sole Facebook presence.

"Anybody who thinks they're going to build ten thousand fans on Facebook and be able to reach those fans without Facebook dipping in your wallet, well they are wrong," says Comm. "It just it doesn't work that way.

"Let's say you've got ten thousand on your fan page. On a good day you may reach a thousand or maybe two thousand—that's on a really good day. Otherwise, you have to pay to promote your post.

"So, you've already worked to build the fans. People who say they like you, want to hear from you, but then you have to pay to reach them, and I think eventually that's going to spill over to personal pages as well." Comm predicts.

How To Bust Through

Think about the purpose of your Facebook page. I suggest considering your Facebook page as a showcase for your best content, as well as good content you find on other's pages. Curating others and posting your own compelling content is a credibility builder.

When you curate content, you can add value by commenting on it. Say why you are sharing it, what you want to highlight about it, and/or give kudos to the author.

If you're simply posting sales messages, chances are you will turn off your audience.

Facebook Action Guide

1. Optimize your business pages and profile
2. Keep your ideal client profile handy and refer to it when accepting Facebook friend request and requesting new Facebook friends.
3. Create an ideal strategic alliance profile and have handy and refer to it when accepting Facebook friend request and requesting new Facebook friends.
4. Set aside 10 minutes each week to review friend requests and either accept, ignore, or follow them
5. Connect with the people you know and meet at conferences, networking groups, community, charity and social events.
6. Identify what is your market craving. Given what they are craving, find and share or develop and post

that kind of content. Ideally publish it originally on your blog and share to Facebook and other networks. This is an easy item to outsource.

7. Reach out and engage with friends and colleagues pages. Comment, Like and Share their content.

8. Watch your colleagues, stay updated with trends (right side of Home page) for new fresh ideas and to stay engaged in the conversation.

Facebook Advertising Myths

The age of paid social media has arrived! Because of Facebook's ever changing news feed algorithm, your Facebook content visibility is declining. That means unless you're great at creating viral content and have a loyal tribe, the majority of your followers will NOT see your posts.

However, paying for Facebook ads can increase that number. The more experienced you are with using the Facebook advertising platform, the more results you'll get.

Myth: Paid Social Media Is Too Costly And Doesn't Produce Sales

"I believe social media is going to be increasingly pay to play, and the price of advertising is going to go up," Schaffer estimates. "Then effectiveness is going to go down. There's never a better time to advertise.

"Invest twenty, fifty, or a hundred dollars," he suggests. "I think every company—if it's going to do social media marketing—should have a paid social part of its budget. Whether you do that on Facebook or Twitter or LinkedIn, there's a lot of ways to splice or dice a budget.

"If I want to try to get results from Facebook, Facebook advertising will definitely be a part," Schaffer explains. "I'm not saying one hundred percent, but definitely a part of what I will do. Buying 'likes' builds fresh blood on your fan pages.

"They are relevant," he adds. "The people that just joined your page will have a little higher affinity than older fans who never engage with your page. It's different than buying irrelevant fans who will never visit your page—ever."

Busted

Your ads don't need to be perfect. You simply need to test and tweak them to determine what works. Facebook Ad expert Nicole Jolie works regularly with clients to bring in sales.

"One of my favorite clients had a video offer that didn't resonate with a large audience," Jolie shares. "We did advertising with and without the video. The split testing proved that the one without video did really well.

"Over eight to ten days that ad netted more than twenty thousand dollars, and by the end of the year it netted her sixty thousand dollars," Jolie reveals. "She got some great value and great clients! So she doesn't advertise as much now—she works on building her business by keeping her email marketing in alignment with the message with which she initially attracted them.

"For Facebook ads to work, the funnel in the back end must be really strong," Jolie admits. "If you walk into advertising simply expecting sales, your buyers won't be interested. If you approach advertising as a means to be the best you can be, give your best service and best benefits, potential buyers will gravitate toward your offer.

"When you understand your potential market, then you can create ads that engage them," she says. "Give something first. Whether it's a coupon, a free download about what a product or service can do for you—give something away free. That will build your email marketing list.

"Then give them something to do, like go check your email right now" encourages Jolie. "Take them through a process that always allows them to engage with you. If you can provide the stories, benefits and free gifts, you can overcome the myths about Facebook Ads and list building."

How To Bust Through

You don't have to spend a lot of money to get results if you know what you're doing. Over time you may find you want to invest more because you're seeing benefits.

"The first thing you want to target is the demographic. Some of the demographics for your ideal client most likely would be age, gender, location, some of those kind of overarching demographics," notes Vahl.

"You also can target based on interest, and the interests are the precise keywords that people might have in their profiles based on things that they've liked, based on things that they said. For example, I like wine so I see a lot of wine ads on Facebook, and so you can use that targeting.

"The place to start would be to simply create a basic ad using the interests and using maybe some other basic tools in the ads manager area such as website conversion so you know how many people are actually opting in using your ad," Vahl explains. "You can gauge exactly how much each lead or sale is costing you. You just put a little code on your website and then actually track how well people are opting in. It's very cool."

As you may begin to realize, placing Facebook Ads requires learning a new set of tools on social media. There are a lot of options. The better you understand them the more effective you'll be, and the less wasted dollars you will spend.

Once you get the hang of Facebook advertising, "you could also use really cool targeting where you can take your email list, upload it to Facebook, and then target specifically all the people on your email list whose email matches their Facebook login," Vahl suggests.

"And from there you can even create this 'lookalike' audience where you then say, okay here are my core people who are on my email list, go out and find a million other people who are like these people, and Facebook will create this lookalike audience and you can make it a larger size or smaller size.

"Usually about a million is where it comes out to be and so then you've got a million potential names you can then add!" Vahl enthuses. "You can target interests on top of that, and really hone down to the perfect potential customers who you can put your ads in front of—it's really powerful!"

"The other really cool thing you can do is this website custom audiences where you put a little piece of code—which Facebook will supply for you—on your website page and then you can have

Facebook build an audience based upon your website visitors that are also logged into Facebook at the time," she adds.

"We built the audience a little bit slowly. But you can then advertise to these people and know that they've come to your website, "advises Vahl. "You can also create a lookalike audience from your website visitors, so it's super cool—there's lots of really cool tools in there that you can use now."

Using **lookalike audiences** and **custom audiences** are definitely advanced marketing techniques. They require some website savvy as well as Facebook power editor experience.

"You can do Facebook ads for twenty-five dollars," Jolie recommends. "That would be five days at five dollars per day and market to ten thousand people. Give them a free five to ten page report.

"When you start an ad on Facebook, however, don't bid them low," Jolie advises. "Bid them super high—one hundred and fifty dollars a day. Watch the ad for the first few hours after it's approved. See how well the click through rate is. Make sure your conversion tracking code is on your site to see how many leads you're getting. Then gradually lower the bid from one hundred and fifty dollars to five dollars throughout the day.

Jolie shares an example of what to use as the subject of an advertisement:

"A realtor might give away a Los Gatos report with the ten best schools and criteria to get into those schools. Target the ad to people looking for a home in Los Gatos. The more clicks the ad gets the cheaper the *cost per click* becomes. Once the ad starts performing, that's when you can pull back the ad spend because your offer is getting a lot of clicks."

Myth: Using Facebook Ads To Get "Likes" Is Going to Get Me Sales

Previously we debunked the myth that having 10,000 "likes" will transfer into sales. The question many businesses ask at this point is: "if getting "likes" doesn't necessarily lead to sales, should I invest in getting "likes" on my business page?"

The answer is—*it depends...*

If you're not going to invest any significant time to create great and engaging content, it may not make sense to build your audience there. Because once you get the "likes," if your content isn't captivating your followers, then they will never visit again. They may even "un-like" your page.

Busted

However, if you're committed to engaging your followers, there are a couple of benefits to building "likes." First, when you get "likes," you get visibility. Your page shows up when someone "likes" another page as "pages you might 'like' also." When you do "like" ads, you can get in front of your targeted niche and their friends.

"'Likes' are great for social proof," Jolie says. "They are not an ad to buy your product.

"For example, Bret Gregory, creator of the *Costa Rica Yoga* Facebook page, has over one million 'likes.' It's a popular page because he ran' like' ads to get social proof. Many are not buyers—one million people are *not* rushing to his Costa Rica yoga retreat. However, as his social proof builds up, it shows up in suggested pages to 'like.'

"It's important to keep going toward a 'like' goal, as well as bring in value on the page. That means it all works together where you have really good content, really good calls to action, and a great following," Jolie notes.

"Joann Darby of *The Girlfriend Life* does a great job of engaging and growing her following. She posts sixteen to twenty-five times a day on her Facebook page, and her followers love it because they feel like they're talking to her. She's brought in people who want to be one of the girlfriends," Jolie reports.

"When you learn how to write and communicate in such a way that it brings people in, everyone wants to be a part of what you're up to. Joann has made money along the way from sharing other's products to offering her own events. So money can be made after you build your community and as long you continue to engage them.

"Then when you do a good job communicating with them and understand what they want to buy, make offers via email and perhaps some by Facebook, you'll get sales," Jolie concludes.

Myth: A Great Way To Promote An Event, Kickstarter Campaign, Or Fundraiser Is Through Facebook Ads

If you've never done any Facebook advertising, Facebook may have sent you a $50 coupon toward Facebook Ads. Upon receipt of this coupon, many newbies to Facebook Ads get very excited and, since there's a deadline, rush to buy ads to promote their latest product, cause, or campaign.

Although promoting a one-off campaign for a brand new product or initiating a Kickstarter campaign sound like great ideas, they can often lead to disappointment.

Busted

Kickstarter campaigns, in particular, require quite a bit of public relations planning to garner attention and achieve attraction. That means promoting not only on social media, but also through your other media coverage, including personal contacts, videos, email list distributions, and personal appearances. If you rely upon Facebook ads only, then $50 isn't going to get you far.

Also if you wait to begin your promotion until the launch of your Kickstarter campaign, you're too late. Facebook "Like" ads can be very useful *prior* to your launch. The three months prior to launch is a very exciting time. Usually you have a lot going on.

IF you're willing to divulge some behind the scenes happenings about your project, then you can generate a buzz and build up the anticipation prior to your launch. This is a great time to build up your followers. Then upon launch, kick off some ads to your followers and you'll have a much greater chance of receiving funding (or sales) then if they see the ad cold.

Facebook Ad Action Guide

1. Develop your compelling give away and overall engaging content strategy
2. Make sure your marketing funnel is built on your back end (email system, messages, follow up)
3. Get familiar with the Ads Manager
4. Do your homework about different types of ads
5. Start with social proof ads to get "likes" and build your familiarity with placing ads
6. Learn the Power Editor (Andrea Vahl has a great blog post about the differences between Ad Manager and Power Editor!)
7. Put the pixel code on your website opt-in page to track conversions
8. Put the *lookalike* audience code on your website (if you get significant visitors)
9. Test and tweak ads

GOOGLE PLUS MYTHS

Google Plus: Just Another Social Network For Which I Don't Have Time

Google Plus, which arrived on the scene in 2011, is a newer social entity. You may have heard rumors that Google Plus is going away, but in case you haven't gotten on the bandwagon yet, it's not going anywhere as far as any of us social media professionals can tell.

What makes Google Plus unique is that it's a social layer within a complex array of Google utilities/programs. In other words, Google Plus is way more than a stand-alone network where you exchange images, videos, and/or text with people. It also involves real time video conferencing via Google Hangouts and its Hangouts on Air (HOA) function.

Hangouts on Air are recordable, and upload automatically to a connected YouTube account. YouTube is another Google-owned property. If you want to create an event and invite people who follow (circle) you on Google, your calendar item will show up in their Google calendar.

"Thus, Google Plus is *like* a social network, but has bigger implications and a much wider reach," explains Carol Dodsley, known as the "G+How2Girl ,"and #1 bestselling author of *A DivaPreneurs Guide to Google+*.

Busted

You may be thinking *what's the big deal about this social layer?*

Well, in addition to all of the above integrations, Google Plus is tied into its search engine, and businesses should take note of this.

If you think about it, Google is what billions of people use monthly for their online searches. When people type into their computer who or what they are searching for—which could be a service or product just like yours—it's the dream of every business to come up in that search!

Actively using Google Plus most definitely enhances your ability to show up in these searches.

In fact, "it's a social layer within not one, but the two top search engines: Google and YouTube," according to Elaine Lindsey, Google Plus evangelist, social media consultant, and host of *Business Banter Plus TV.*

Interested now? Given that Google Plus is a newer entity, your fellow business colleagues may not be taking advantage of it—and that means you have an incredible opportunity to stand out!

How To Break Through

Everything we do in our online business development is about getting found. Why not start from search?

YouTube is a part of the Google family, and ties video into a search. Lindsay encourages us to use Google Plus "as a great place to jump into social and go from there. When you talk about small business, the greatest leveler is having Google in your pocket. When you add authorship and the various pieces of your Google Plus account, you start to get your social activity to show up on search."

The real attraction to Google Plus is their Hangouts on Air (HOA) tool.

"People can use HOA not just for prospects, not just to get the message out, but as a great way to give kudos to your clients," Lindsay explains. "You can help your customers who have questions by doing an interactive hangout. This gives them a chance to talk to you directly."

Many social media people resistant to Google Plus have been won over by the Hangouts On Air tool.

"When I was in the process of launching a new brand, I realized I was being a wimp about it," reveals Brandee Sweesy—a leading Hangout marketing expert with 20 years marketing and public speaking experience—about how she got to be an expert.

"Since I'm deadline driven and work well under pressure, I gave myself a deadline forty-eight

hours before my birthday launch with a hangout on my birthday. I knew nothing about Hangouts! I thought it would be just like a Skype call. I did thirty-six hours of organic promotion posting to social media channels. I invited a few guests to be on, and got seven hundred viewers on my first tries.

"I looked around and thought: *why aren't we using hangouts all the time?* However, everything didn't go exactly according to plan, so I hired an expert to teach me the tech side. I started focusing on the marketing and strategy—my first goal was to simplify the whole process because I'm not techy," says Sweesy.

Though this book has not delved into YouTube as a social network and the implications of video, all references to content include the use of video. Video is important and easily created if you take advantage of Hangouts On Air. You can produce videos with yourself on camera, or as a voice over the screen sharing of a PowerPoint presentation.

Myth: No One is Using Google Plus— It's a Ghost Town

Rumors swept like wildfire about Google Plus as people complained about this latecomer network. They started out saying no one needs another social network. And, for those already challenged by social media, resistance to Google Plus became stronger than ever. Many people declared that "no one is on it" because they didn't find their friends using it.

Busted

The reality is that there are hundreds of millions of people already there. Plus, millions more are poised to activate their Google Plus profile with one simple click from their gmail account.

> **"Google Plus is actually only a ghost town when you don't know how to use it." ~ Carol Dodsley**

Google Plus trainer Dodsley explains "Google Plus doesn't work like Facebook. When you circle someone, you are simply saying you want to see what they post. When you share your own posts with your circles, the only people who will see it are those who have circled you back.

"You have to share your posts publicly to start with, and then more people begin to see you and to circle you. When someone circles you, they are essentially saying, 'I like you. I like your stuff. I want to see more of it.' And all of a sudden the ghost town disappears."

Since Google Plus is a social layer and tied to Google search, let's look at it in a different light:

"If you're in a business and people want to find you, then they're actually in a buying mood," notes Dodsley. "Do they go to Facebook and Twitter? They might ask their friends 'do you know where there's a decent plumber?' But most of the time they Google it if they're seriously looking for something.

"Everything you post publicly on Google Plus is indexed for search," she explains. "You can link your Google Plus profile with your website, which helps you get indexed even faster for search. You can embed posts from Google Plus into your website, thus creating links all the time.

"And, everybody you start to connect to and engage with on Google Plus is helping your visibility on Google. Google is where people go when they're looking for stuff. So this where most people say 'but there's nobody on Google Plus that I know.'

"You don't need to know anybody on Google Plus! You'll get to know new people! Don't try and bring your current friends over because they're quite happy in the bar. You can go back to the bar and have a chat. But come and meet new people in the local bar or the new modern bar down the road as well. Remember it's not the people that you're talking to necessarily, it's the people *who are looking for what you do and who love googling for it*. You need to be found by them!

"Look for and circle influencers in a 'go to give' way," Dodsley advises. "Go and find the influencers in your niche and share their great content with other people. That makes your audience see you and think' oh, this is pretty neat' because *they're* not doing all that research and bringing out all these great posts.

"So, they'll think: *I need to follow this person because she's obviously following—or in connection with—other people who are posting great stuff that I'm really interested in, and they're doing my job for me!'*

"Also by following influencers in your niche, you'll be seeing who's following them," Dodley suggests. "You'll be able to join in the conversation. I joined communities not to show my stuff but to add value. And the more I did that, the more people started circling and noticing me. "

"Everything public on Google Plus is indexed for search." ~ Carol Dodsley

Myth: I Don't Want to Engage As My Personal Profile—I Want To Engage As My Business Page

On Google Plus, people are primarily engaged in creating relationships through their personal profiles.

"It's not *your shop* that goes networking, it's *you*," Dodsley explains. "It's not your office that walks into a network event and says 'hello,' *it's you*. It's the same online.

"Yes, you should have a business page because that's where you will promote your business, and share your offers, sales, and other things that are of interest to people regarding your business.

"Again, this is a slight mindset shift from other social networks," points out Dodsley. "I would say build your personal profile on Google Plus, and if you're going to build your business page, be prepared to put a lot more work into it. And make sure that you've got a real personal sort of image on that business page because a lot of people don't like talking to logos," she advises.

Google Plus Action Guide

1. Optimize your profile
2. Link your website to Google Plus
3. Circle Influencers and share their great content publically
4. Post publically
5. Contribute to communities
6. Start a Hangout TV Show and create a circle of people who want to be invited

LINKEDIN MYTHS

Myth: The Bigger Your LinkedIn Network the Better

Are you playing the "collecting connections" game on LinkedIn? Having lots of connections feels good, looks good, and makes you more visible to an exponential number of people. However—be forewarned—the larger your network on LinkedIn, the more trouble it may cause you.

I must confess that I have hundreds of people who've requested to connect with me on LinkedIn, but I haven't yet accepted them. You may wonder *why* this is so...

Early on, I realized LinkedIn is *the* social networking site designed specifically for business networking. Unlike Facebook, Twitter, and Instagram—which are socially-oriented *first,* business-oriented *second*—people go to LinkedIn for business purposes *only.*

So, if I accept all those requests, people will most likely want to engage with me. As a busy professional, I can develop only a handful of new business relationships at a time.

Busted

"Where LinkedIn really shines is in the one-to-one relationships, and helping to nurture that one-on-one relationship, explains LinkedIn expert von Rosen. "Yes, having thirty thousand connections gets me on Forbes, but it's hard to manage a network that size. I have to use all kinds of secondary tools to manage a network that size.

"The benefit of having a more strategic network, and of keeping up with your connections in that more strategic network, is less work and higher reward," she adds. "By the same token, If your network's too small and too strategic, you're won't gain visibility. Then it becomes simply a place to keep up with your current clients, instead of gaining new ones, so there's really a happy medium between the two."

How To Bust Through

When I posed the question to Melonie Dodaro about what to do with my hundreds of LinkedIn connections requests, Canada's LinkedIn Expert and author of *The LinkedIn Code*, busted this myth for me with a simply strategy:

"When you have a bunch of connection requests piled up, you need to have a systemized plan in place that minimizes the time it takes," Dodaro noted. "You can personalize the top of the message with your recipient's name, and copy and paste the follow up message. You may not want to ask them a question, as that may cause too many responses for which you won't have the time to engage. Instead, provide something of value in a way that gets them on your email list."

Myth: Joining Groups In My Industry Is Great Way To Solicit Clients

One of LinkedIn's best engagement opportunities occurs within groups. You may join up to 50 groups on LinkedIn. Being a member of a LinkedIn group generally allows you to request to connect to anyone in the group. In addition, the discussion area is where you can network and contribute your value.

Busted

However, it's easy to misuse groups and get too little traction.

"The biggest mistake occurs when you join only industry specific groups, instead of also including groups where your target market interacts," Dodaro points out. "There's nothing wrong with joining industry specific groups—you *should* join a few to stay up to date, know what your peers are doing, and monitor competitors. However, if you don't sell to your competitors, there's little else you can do there.

"You can also join groups related to your personal/business interests where you can learn sales, social media, etc.," Dodaro advises. "The vast majority of LinkedIn groups you'll want to join are those connected with your target market. It's in those groups you can start discussions, and engage in others' discussion comments. Make sure you have a very structured lead generation plan on LinkedIn."

How To Bust Through

"Lay the foundation with a client-focused profile," she suggests. "Then make sure you understand LinkedIn etiquette and best practices, and create a social selling plan. That includes prospecting and finding ideal clients through advanced search and LinkedIn groups.

"Connect with them via a personalized connection request message, *and* connect with them afterward. That way you'll create relationships with your new connections, instead of simply collecting them."

Also, be careful on groups with your engagement. Spend some time reading *only* and noticing what others are asking and sharing. Furthermore, notice what types of topics attract the most activity. Determine if people share their blog content or others' content freely. Is promotion acceptable or not? There is the "promotions" tab you can check out.

By far, the best way to engage on Linked In groups is through discussion dialogs. Then you can invite those with whom you want to connect more meaningfully. Take the conversation to the next level with a message to them.

Myth: Put Keywords In Your Profile Name

Many people have been told by so called "experts" to optimize their LinkedIn profile by including keywords in the name field.

Busted

Whoops! The problem with this is that it's against LinkedIn's end user agreement. So if you've placed keywords, symbols, phone numbers, and/or email addresses in your profile name field, you could be blacklisted by LinkedIn.

Change your profile back to your name *only* as soon as you read this! Otherwise, LinkedIn will essentially hide your profile from being found by their internal search!

Von Rosen learned that buying into this myth caused visibility problems for her early on.

"I see a lot of people putting email addresses, phone numbers—and guru of this, expert at that—in their last name field. This goes against LinkedIn's end user agreement, and LinkedIn will blacklist you if you do that," von Rosen cautions.

"I did it because someone told me to, and it did help with Google results. But I went from getting between twenty to thirty warm leads a week to three per week on LinkedIn. And I'm still blacklisted! I do show up under my name now, but only because I'm everywhere else on the web."

Myth: Endorsements From People You Don't Know Are Not Good

There is a difference between LinkedIn *endorsements* and LinkedIn *recommendations.* Endorsements are fairly intrusive because they show up on the top of every profile to which you go. The initial reaction of many was annoyance at LinkedIn's persistent request on every profile.

Busted

Endorsements are similar to Facebook "likes." Every click is a way of saying "I like this person for this category." It looks good to have a lot of endorsements. Plus, people like to endorse you. It feels good to receive an endorsement, and others get visibility on your profile simply for recognizing you. There's no downside to accepting endorsements unless you don't get many.

"LinkedIn endorsements are like a Klout score," explains von Rosen. "People *do* look at these numbers, and if you've gathered twenty or fewer in your area of expertise, it doesn't look very good. Accept any endorsement you get because some people take them seriously!"

How To Bust Trough

Give it a go—endorse people for the skills you know are in their scope of business. Some skills are not relevant to that person, so you may want to go directly to their LinkedIn profile and select the specific skills for which you want to recognize them.

"Endorsements are a great way to engage," von Rosen continues. "When someone endorses you, that's a great opportunity to message them and say, 'thanks for the endorsement, I really

SPECIFIC SOCIAL NETWORK MYTHS ▪ 131

appreciate it, I'm going in and endorse you for this and this and this now, or how have you been or how are the kids,' or whatever. It's a talking point. The more real life, face-to-face, or online talking points we can get, the better. Endorsements are a great way to engage with your audiences!"

Myth: Posting Your Webinars in Group Discussions Garners Good Results

While it's tempting to join a group, and start sharing your webinars, free offers, or paid events, stop and think before you post. How you would feel if you were another member seeing your posts, and/or the group owner? Do you think posting your promotional content will be well received?

Busted

If you do put out too many promotions in the discussion area, you can get "site wide auto moderated," or SWAMed. Group members can report your discussion as a promotion, and if too many reports are received you'll be SWAMed.

"If you get SWAMed any time you post something, it has to be moderated by the group moderator," von Rosen elaborates. "If the group moderator is like me, I may go in once or twice a week, see what's in my moderation queue, and approve or disapprove. Of course, if it's a webinar for sale, I'll either put it in promotions or delete it. Once you've been SWAMmed, you're on LinkedIn's radar as suspect," she warns. In other words, LinkedIn may restrict your account at their discretion.

How To Bust Through

"Remember people, it's social!" implores Business Systems and Hardwiring expert Lisa Mininni. "Even though savvy people know business is built via relationships, they forget that the minute they start posting. Use a strategic approach to social media instead," she advises.

"Connect with people, be yourself, and then focus on your

preferred client profile. Otherwise, it will absorb too much of your time."

If the group in which you're participating is not a promotion-oriented group, be very careful NOT to promote your webinars/events in the discussion area. Put your event in the promotion area.

Or, von Rosen suggests: "Ask questions about it. Say, hey, I'm doing this webinar and have a couple questions. I'd love to get your feedback! Also, please feel free to sign up. Can you tell me how often you use LinkedIn, do you pay for LinkedIn, what LinkedIn services do you use?

"Positioned this way you're simply asking the group for help and/or input, and relegating promotion to an aside. If it's an outright promotion then put it in promotions."

Myth: Email LinkedIn Connections with Promotions About Your Webinar

Many people have started to use LinkedIn messaging as an event and product promotion tool. Maybe you've seen the messages like this one: "Given your background, I thought you'd be interested in this _____ "(fill in the blank with webinar, investment opportunity, product, etc.).

Busted

To be clear, emailing your LinkedIn connections with a promotion about your upcoming webinar is considered SPAM—if the connection doesn't know you. Be VERY careful about sending out this kind of the email, as it's relatively easy for a user to mark your message as spam.

Most likely, LinkedIn *will* crack down on your email if too many people start getting reported as using the system to promote their business, product specials, and services.

"The more people do this, the more likely LinkedIn will come out with something like SWAM, " von Rosen predicts. "Then you won't be able to send messages on LinkedIn. It's bad in LinkedIn and getting worse. It never used to be that way."

Use some best practice smarts with LinkedIn, and only email the people who you *know* want to receive these kinds of email.

Myth: Being a LION (LinkedIn Open Networker) is Valuable

A LION is someone who accepts *all* LinkedIn invitations. This started early on and used to be a recommended way to get more connections (if that was your goal.)

Busted

LIONs are mostly known today as spammers. It's no longer a good thing to accept invitations from LIONs. LinkedIn limits the number of invitations you can accept to 30,000. LinkedIn also limits the number of invitations you can make to 3,000.

In general, it's a good idea to accept invitations from people you know as well as from those with whom you can strategically align for business purposes.

How To Bust Through

"If you receive an invitation from someone you don't know, look at their profile first," von Rosen recommends. "If they fall in the category of a good client, prospect, or partner for you, then accept their invitation. Segment them by tagging them, and then begin a relationship with them. Don't just accept invitations for the sake of accepting invitations," she adds.

"Also, please take the word LION out of your last name field, out of your professional headline, and out of your entire profile copy. If you're sharing LION images, stop doing that. You're only positioning yourself as a spammer," von Rosen cautions.

Myth: B2C Companies Cannot Effectively Use LinkedIn

Businesses who serve end customers, *a.k.a.,* B2C companies, are often discouraged from using social media to connect with end clients.

"If you have professional services like website design, accounting, law, coaching, consulting, marketing, and selling to businesses, LinkedIn is great for your lead generation and social selling," according to Dodaro.

LinkedIn is set up to allow you to search for people in business, rather than via their interests and challenges. Facebook, Twitter, Pinterest, and Instagram are better for those consumer-oriented searches. Considering this, should the B2C companies step aside on LinkedIn?

Busted

If you serve end consumers primarily, your goal on other networks may be to seek out customers directly. However, consider a different but potentially extremely effective strategy for finding strategic alliances on LinkedIn.

"Financial advisors who serve the end client can use LinkedIn well if they focus on finding other professionals who serve the same audiences, " says Dodaro. "Those might be accountants, lawyers, and various insurance agents. To use LinkedIn well you must understand where you're going to get the best leverage.

"Another example would be if I'm working with a restaurant," she continues. "I'd verify that they have a catering aspect of their business, and then suggest they connect with companies that do corporate lunches. If you're a massage therapist, think about strategic alliances such as chiropractors and other wellness industry companies. Understand who your target market is and how you can best reach them through LinkedIn. You can be more creative in this B2B space!"

How To Bust Through

Think about what types of businesses also serve your clients. If nothing comes to mind, focus on one or two of your ideal clients. Where do they shop? Who do they go to for advice? Where do they spend their leisure time? What products are they spending their money on? Then search LinkedIn for companies and professionals who may want to collaborate in some way.

For example, I help rising entrepreneurs use social media for business growth. It may be arduous to figure out who is a rising entrepreneur, but if I align with leading entrepreneur mentors and coaches, I usually find a way to joint venture with them because they are rarely providing the kind of effective social media support and training to their clients that I do.

I initiate a relationship on LinkedIn with a simple invitation, and ask them about their business. Eventually, if the conversation is going well, I'll invite them to speak on the phone. From there it may go in several different directions.

The key is to be clear about the intention of this call, and what you want overall from this relationship. My intention is to provide value, increase my connection's value to his or her clients, and determine if we have aligned businesses. In other words, I want to determine whether or not this customer's clients are similar to mine.

Be sure to create your LinkedIn company page and showcase pages. When LinkedIn eliminated it products and services pages, it introduced showcase pages, but these two are not the same.

"A showcase page is more like a sub-page," explains von Rosen. "Products and services pages were pretty stagnant and, other than being able to play video, they were really more of a brochure. Showcase pages allow you to disseminate content and market to your audience.

"What I tell my clients to do first is reserve your showcase pages *now*," she continues. "You're allowed ten, but they each have a unique URL. "

When you create your showcase page, it will look something like this: linkedin.com/company/y*ourshowcasepagename*.

"Get your URL's now," von Rosen emphasizes. "A lot of them are gone—*internet marketing* is gone, *LinkedIn marketing* is gone (because I got it!)

"Think about it like a Twitter page," adds von Rosen. "Really know the particular audience for whom you want to create your showcase page—know their wants, their needs, and then provide them that content. Once in a while you can throw in your own product or service. It's about creating followers, just like you would on Twitter.

"There's a big nice hero image similar to Facebook. I advise people to place your USP (unique sales proposition) in the description area, and then to add content you want to share to your unique audience. I'm excited about showcase pages!

"LinkedIn's redone their inbox, so now when you click on a person's name you can see a digest of your communication with them and you can set up a reminder. If you don't want to respond to them right away you can setup a reminder and you can write notes on the person," von Rosen adds.

"LinkedIn introduced *LinkedIn publisher* about which I am really excited, mainly because it's the same thing that famous influencers got to share blogs like the heads of states, company leaders, and digital media leaders like Dave Kerpen, NY Times best-selling author of Likeable Social Media and chairman of Likeable Media.

"Those folks originally got the ability to create blog post length posts, and now we get that, too. We actually have that publisher ability now. These LinkedIn blog posts searchable by the article search by keywords so that is absolutely huge. They get enormous visibility, not only on LinkedIn but I'm guessing pretty soon—because Google and LinkedIn really like each other—they're going to have that connection there, too, we'll see. Also you can share the LinkedIn publisher blog posts into your groups."

"LinkedIn taketh away but LinkedIn giveth to us as well."
~ Viveka von Rosen

Myth: Adding Connections from Your Email Is A Good Way To Get More Connections

Since LinkedIn limits you to 3,000 invitations, you may NOT want to let LinkedIn connect to your email service. These 3,000 invites include *all* the people they will automatically invite if you select LinkedIn's option to send out invitations to your Gmail, Hotmail, Yahoo, AOL or other email contacts. If you have close to, or more than, 3,000 contacts, all your invites will be used up with that option.

Be careful not to accidentally do this! LinkedIn makes it looks like a sign in page, but it's actually asking you for your Gmail or Hotmail or AOL (or whichever) account from which it wants to blast all of your connections with invitations to connect with you on LinkedIn.

As a side note, don't create duplicate accounts.

LinkedIn Action Guide for Small Business

There are a great many new features that have become available in LinkedIn at the time of this edition. And it's a great time to take advantage of them.

1. Set up showcase pages with keyword URLs and offer updates to your audience
2. Get in Group digests by contributing to discussions in the early morning (these have been known to show up in the email digests!)
3. Use the publisher to create content 1-3 times a week to start
4. Create your strategic communications plan with new connections

5. Accept connection requests from people who are clients, or who would make a good prospect or partner for your business
6. Send follow up messages
7. When appropriate, take the online connection to the phone or an in-person meeting

Extras:

- You can mention your connections when you post updates
- Hashtags don't really work well yet in LinkedIn
- You can block people from emailing you on LinkedIn

INSTAGRAM MYTHS

Myth: You Can't Build A List Or Sell From Instagram Because You Can't Add Clickable Links To Posts

That's right—Instagram is one of the few social networks that don't make *clickable* links that you can add to descriptions and comments. There's only one place you can add clickable links on Instagram, and that's in the *bio* section of your profile.

We're so used to posting links in the hope of generating traffic from our social networks, that it's become the norm as a way to get traffic. In fact, if you have a Pinterest account and have seen a ton of traffic generated because of the clickable links associated with pins, you may be baffled about how Instagram can possibly work for your business.

Instagram requires a *rethinking of strategy* to build brand awareness, engagement, and new leads.

Busted

The biggest difference between Instagram and other popular networks discussed so far, is that it's primarily a mobile app-based system. **You must use a mobile app to add your posts.** From your computer, you can log on, but you are can only view, like, and comment on people's Instagram posts. You cannot sign up there. You must do that via their mobile app.

To some this may be frustrating, but to those who live on their mobile phones, it's not an issue. Nonetheless, Instagram is growing by leaps and bounds in popularity.

Studies have shown that it has the second highest engagement rate next to Facebook. A significant number of users open their Instagram app more than two times a day. This means you have a great opportunity to get engagement!

Obviously, though, the trend now is mobile. People reach for their mobile device to do everything from set reminders and alarms, find out the weather, text friends, check their investments, sell their wares, buy stuff, swap real time images with family, Facetime/Skype, play games, watch movies, listen to music and podcasts, and the list goes on.

So it makes sense that users engage in social networking on their phones, too.

"I've sold hundreds of my courses and consulting services because of my Instagram activity," she reveals. "People think that because you can't directly click on the picture, it means you can't make sales. Since a lot of people think this, there is less competition and a huge opportunity on Instagram."

The reason I am able to get sales from Instagram: I've already given information that is both personal and valuable. – Tar'Lese Rideaux

How To Bust Through

Building relationships on Instagram is about creating that two-way dialog. Posting images with descriptions that are meaningful to you, and interesting to your audience, is simply the first step.

Tar'Lese Rideaux, a network marketer and Instagram consultant who gets 100% of her business from Instagram and Facebook, offers a unique and effective business development strategy.

If you utilize the direct Instagram features, you can connect one-on-one with your followers, and your ability to convert contacts into clients will skyrocket!" Rideaux claims. "To grow your visibility, start commenting on the Instagram movers and shakers content that your market loves. Only about ten percent of people comment on Instagram posts, so if you start commenting a lot, it makes you stand out."

Also, Rideaux recommends that you optimize your bio section on Instagram with *list-building* calls to action. From time to time, it may make sense to reference the link in your relevant post descriptions. One of Rideaux's clients garnered more than 800 leads in two weeks through her free giveaway by using Instagram for her business.

Instagram is one of the most popular places to use and benefit from hashtags. A *hashtag* is a word with the "# " in front of it. When you place the "#" in front of a word, or series of words combined into one word, you create a link that, when clicked by a user, finds all posts with that hashtag.

For example, if your business is about fashion or hairstyles and you place #fashion or #hairstyles in the description, you have a nice chance of being seen and followed by other users interested in that.

"Hashtags are the keys to your market following you," according to Rideaux. For example, she encourages her network marketing clients to use #networkmarketing #WAHM #MLM. WAHM stands for Work At Home Moms. And MLM is multi-level marketing.

This gets you in front of your market, but hashtags alone won't get you sales. So when posting to Instagram, remember to make 80% of your content personal/valuable content, and only 20% business.

Here's the real gold: use your descriptions on these business posts to invite people to message you "directly." And/or direct message anyone who asks a question about your services in comments. Create an image that says "Let's Talk" or something similar with your number.

Then reference their message and let them know in the description you're looking forward to speaking with them. This is the best way to get in touch directly with your market so you can serve them.

Coach Laura's Breakthrough Questions

- Does your target audience have smartphones?
- Do the research on Instagram—are other influencers in your market connecting there?
- What hashtags are you using?

Myth: You Should Get Automatic Comments And "Likes" On Posts

One of the big myths going around is that getting a ton of auto-mated comments and "likes" is a great way to build your Instagram following and get business. This is another social me-dia status symbol. And once again, this idea is completely false.

Busted

First of all, Instagram doesn't like automation. For that matter, none of the social media networks allow it. Instagram has been very rigorous about how other programs can access and inter-face with their platform. They have shut down access to automating "likes."

The more people try to automate Instagram, the more re-strictive they have become. Scheduling at one time was possible through an app that interfaced with Instagram, but as of this printing, currently that's not even an option.

How To Bust Through

Although there is limited automation available for Instagram, that doesn't mean you should shy away. Instead, that's the good news! There is less possibility of spam, so hopefully it will re-main a fairly clean network. If you do see any "spam" posts, un-follow the user and report the content as inappropriate. You can do that directly from the post.

In case you're worried that Instagram will take up too much of your time, rest assured.

"I'm very strict with my time—I value my time and how much is my hour worth," says Rideaux. " "I realize that the time on Instagram or Facebook is time could I be spending with a client. So I track my Instagram stats through Iconosquare (an online tool

"I know which days are best to post," she continues. "I have five accounts. **On my personal accounts Wednesday and Thursdays from noon to 2 pm are best for business posts because I get the most clicks on my bio link and direct messages at those times** "Rideaux tracks the activity on her account using Iconosquare. This site allows you to see which posts get the most likes, comments. Use a bit.ly link in the website field to track when you get clicks.

"Then rest of time I spend responding to comments—I stick to days and times when people are buying," Rideaux notes.

Instagram Is For Teenagers And Fashionistas Only

A mobile based app often gets the attention of the younger crowd. On top of that, an image based app attracts pop culture. Does that mean your brand doesn't belong there? Maybe, maybe not.

Busted

A majority of users on Instagram are under the age of 35. This market is young, but they are not simply teenagers. They include the rising workforce. It's worth considering Instagram if you want access to young adults in urban areas.

If your business serves the young adult through coaching, personal development, professional development, fitness services, entertainment, travel, fun, architecture, sports, hobbies, or retail products, you'll want to consider Instagram. Do your research.

How To Bust Through

Begin with a simple search on Instagram for your industry, hobby, product, or service. Check out what people like and if they are following people in your industry. If there's some good activity, chances are you'll get some great traction.

If this network simply doesn't reach your market, or interest you or anyone in your organization, it is fine to back burner it. However, consider seriously whether or not your business is right for Instagram. It's one of the most underutilized networks that's proving profitable when used strategically. This may be your chance to breakthrough and have a great win on social media!

Additionally, there is significance to the fact that Facebook owns Instagram. Their networks are tied together so you can post a video or image to Instagram and then automatically share it to Facebook.

Hosting a contest is a great way to grow your social visibility. If you've wanted to run a contest on social media, Instagram makes it pretty easy and actually gives you tips and case studies to help you get started https://help.instagram.com/464700830247492

If you have a growing active audience on Facebook, you can announce your contest there. Then you might even consider creating some Facebook ads to promote your contest.

Instagram Action Guide

1. Optimize your bio with a call to action for example, ♥ Click link to get Savvy Social Media Marketing Blueprint ♥ #FREE guide: and in the website field include the link where they can receive it.
2. Spend time researching your keyword hashtags
3. Comment on others' posts that use your hashtags
4. Create images and plan your descriptions
5. Plan a contest and promote it

AFTERWORD

We have been on quite a journey together throughout the book. You've conquered myths and hopefully you are more energized and excited about using social media for your business than ever. With your new perspective, you can choose the networks you want to shine in.

To stay motivated and connected I want to underscore that our relationship has only begun. As you access the SocialMediaMythsBusted.com blog you can join me and each of the myth-busters in up front and personal video interviews. Each video interview is housed in a blog post and there is so much more offered there than I could have possibly put into the book. So visit our blog to see what we're up to and connect with us.

In these recorded interviews, many of our myth-busters admitted to having started from humble beginnings and some even admit to being challenged on the networks themselves. We're all human and each has our hurdles to overcome. The thing about these hurdles on social media is that they are easily surmounted. Most of the time all it takes is a mere simple mindset shift and a small amount of time. This is not rocket science. Your unique genius is more than enough to thrive on social media.

If our myth-busters can do it, so can you. For example, Julie Renee wouldn't even go online to any website six years ago for fear that her computer would get a virus. Now she is a highly sought after speaker, healer and celebrity on Facebook.

Melonie Dodaro considers herself a slow learner and not techy at all. In fact, her cousin had to walk her through setting up a Facebook account. Now she is considered Canada's LinkedIn Expert and author of The LinkedIn Code.

Tar'Lese Rideaux was didn't want to get into social media but made herself learn it. Now 100% of her income comes from the leads she gets from Facebook and Instagram.

Nicole Jolie was a triathlete coach and not techy at all. But started using Facebook with her team and without really knowing much she got heaps of attention, free racing entries, and swag. Now she is a Facebook Ads genius.

Brandee Sweesy admits to hardly posting on Google Plus, yet rockstar internet marketers hire her to train them on how to use Google Hangouts On Air to build their email lists.

Natalie Ledwell , Founder of Mind Movies, LLC who now has an email list of over 600,000, could not comprehend why people would want to share details about their life on a page for other people to look at. Perhaps you even ask yourself like she did "why would anyone want to do that?" Today Natalie is a huge advocate because as she says, "Through social media you get relationships and connections with your community you can't get through email. I get approached to speak on tele-summits and in person speaking engagements."

There is no accident that our myth-busters achieved success on social media. Social media is simply about bringing the humanness into computing and our online lives.

Humans crave connection. And social media fills that need in one of the most revolutionary ways ever invented. Yes that may leave us open to misusers. But through education, good practices and myth busting, social media can be used for good. The more people using it for good the greater impact we can have to making the world a safer, peaceful, inspiring and thriving place for all.

ACKNOWLEDGEMENTS

Meet Our Myth Busters

I offer my humble and deep appreciation for all of our myth busters. This book is enriched because of their contribution.

Take to heart and put into action the insights, tips and tools they have recommended and woven throughout the book for they are golden.

There were even more enriching insights that each myth-buster shared during our interviews. Luckily because of social media, we recorded most of these interviews using Google hangouts.

The original live interviews are available on SocialMedia MythsBusted.com blog.

Here is a little highlight about each myth-buster and where you can find more about them.

Special thanks to **Joel Comm** for generously writing the Foreword and for being an encourager and mentor from the start.

Our myth-busters featured at publishing include:

Joel Comm: Web Pioneer, Serial Entrepreneur, NY Times Best Seller, International speaker, Consultant (JoelComm.com)

Jenn August: An internationally recognized Speaker, the Success Mindset Expert and Founder of a global phone networking organization Women's Success Tribe, where women create profitable business connections that last! (JennAugust.com)

Larry Benet: World renowned connector and founder of SANG Events and keynote speaker. (LarryBenet.com)

Deb Cole: Authored 1st Book on Twitter: Twitter Revolution, New Media Consultant, News Go-to-Girl, Marketing Director for New Media Expo (@NMX), speaker and unofficial racecar driver. (CoachDeb.tv)

Emmanuel Dagher: #1 International Best Selling Author, Expansion Catalyst, Transformation Specialist, Intuitive & Humanitarian whose priority is to reconnect those who are ready back to their greatest potential. (EmmanuelDagher.com)

Tina Dietz MS, NCC: International small business and lifestyle design expert known as "The Voice 10,000 Thriving Businesses." (ThisIsTinaDietz.com)

Teresa de Grosbois: International speaker, 3X bestselling author and influence expert sought by entrepreneurs and writers wanting to better understand how local word of mouth can suddenly turn epidemic. Founder of the Evolutionary Business Council. (WildfireWS.com)

Karen Dietz: Business story-telling speaker and consultant. Author of *Business Storytelling For Dummies*. Top global curator in business storytelling. (JustStoryIt.com)

Melonie Dodaro: Bestselling author of *The LinkedIn Code* and dubbed by the media as Canada's #1 LinkedIn Expert. Top 10 Social Media Blogs by Social Media Examiner helping businesses & sales teams leverage LinkedIn & social media marketing. (TopDogSocialMedia.com)

Carol Dodsley: Also known as the G+How2Girl, leading Google+ and Hangouts Trainer, consultant and Google+ top contributor. Creator of GPlus and Hangout Professionals and author of the #1 best-selling book A DivaPreneurs guide to Google+. (CarolDodsley.com)

Denise Gosnell: Author of 8 business and technology books, attorney and member of the bar of Indiana and the United States Patent & Trademark Office. She focuses on serving publishing, technology, and related companies on intellectual property, licensing, Internet, and related legal matters. (GosnellAssoc.com)

Stephanie Gunning: bestselling author, editor, and publishing consultant specializing in books on the topics of health, spirituality, self-help, business, memoir, and new thought. (Lincoln SquareBooks.com)

Samantha Hartley: Founder & President of Enlightened Marketing helping experts and entrepreneurs create Jaw-dropping Client-getting Messages™ that attract perfect clients. (EnlightenedMarketing.com)

James Hickey: James Hickey is one of the top leaders in the Internet Marketing industry for Business Owners, Entrepreneurs and Start Up Companies. He also created the Internet Marketing Training Center in January 2011 to train individuals to become Internet Marketing Consultants. (InternetMarketingTrainingCenter.Net)

Nicole Jolie: A new media expert who combines the power of social media with savvy online marketing to create socially smashing results for her clients using the potential of the Internet to deliver messaging and branding that makes an impact. (SociallySmashing.com)

Holly Kolman: Relationship management / Social Media Strategist specializing in Twitter to grow authority and follower numbers for Authors, Speakers, Leaders, Business Owners and Executives so they can attract media opportunities, speaking engagements and sales. (HollyKolman.com)

Natalie Ledwell: Best-selling author, speaker and Life Improvement Crusader and founder Mind Movies LLC (NatalieLedwell.com)

Elaine Lindsay: Google Plus evangelist, social media consultant, speaker and host of Business Banter Plus TV. Dean of Digital Media for Social Buzz U. (TROOLSocial.com)

Gail Martin: Bestselling fiction and non-fiction author with more than 14 books published. international speaker and social media expert. (DreamSpinnerCommunications.com)

Sharon McRill: President of The Betty Brigade a full service relocation, moving coordination, event planning, personal assistance and concierge company. (TheBettyBrigade.com)

Lisa Mininni: Business Systems and Hardwiring Expert, speaker and best-selling author of *Me, Myself and Why* and creator of The Entrepreneurial Edge System. (ExcellerateAssociates.com)

Sierra Modro: Technology Evangelist, speaker, and consultant who brings clarity to complexity. It's Tech Savvy - Simplified. (SierraModro.com)

Julie Renee: America's Brain Rejuvenation Expert, author and survivor of atomic bomb testing and multiple cancers who uses Cellular Quantum Mechanics to regenerate cells, tissues and organs. (JulieRenee.com)

Kathryn Rose: Bestselling author of 9 books, social media strategist and co-founder, Social Buzz Club, LLC (amazon.com/author/katrose)

Tar'Lese Rideaux: Network marketer, Instagram marketing consultant and creator of the Insta-Influence course. (Insta-Influence.com)

Ted Rubin: Social Marketing Strategist, Keynote Speaker, Brand Evangelist and Acting CMO of Brand Innovators. In March 2009 he started using and evangelizing the term ROR, *Return on Relationship* (book published January 2013), hashtag #RonR. (TedRubin.com)

Neal Schaffer: Global social media speaker, author of *Maximize Your Social* and two LinkedIn books, social media strategy consultant, and founder of the social business blog Maximize Social Business. (MaximizeYourSocial.com)

Lisa Steadman: Relationship expert, bestselling author, writer/producer, media personality, and highly sought after voice for women who are redefining what Having It All looks like. (LisaSteadman.com)

Brandee Sweesy: Leading Hangout Marketing Expert with 20 years marketing and public speaking experience she helps thought leaders and businesses to Engage! Ignite! Expand!® their audiences with all forms of Live Video. (HangoutsForBusiness.com)

Andrea Vahl: Social media consultant, speaker, co-author of *Facebook Marketing All-In-One For Dummies* and also double as, Grandma Mary -Social Media Edutainer. (AndreaVahl.com)

Viveka von Rosen: Author of best-selling *LinkedIn Marketing: An Hour A Day* and internationally known as the "LinkedIn Expert". Ranked Forbes 20 most influential women 2011, 2012, 2013 and 2014. (LinkedIntoBusiness.com)

Denise Wakeman: Your Guide to Better Visibility on the Web. Online Business Strategist, Founder of The Blog Squad, Co-Founder of The Future of Ink, and Host of the popular Hangout show Adventures in Visibility. (DeniseWakeman.com)

So many people have contributed to the success of the book. I send my heartfelt appreciation to my clients who have made me better and the colleagues who keep me information.

I have great appreciation to our Editorial Team including Publishing Consultant and myth-buster, Stephanie Gunning.

Huge thank you to our Editor, **Sheri Horn Hasan of** Karmic Evolution for editing the manuscript in record time. Karmic Evolution's mission is to help authors give voice to their vision and inspire their audience through the beauty & power of the written word! (KarmicEvolution.com)

Most of all I give my deepest appreciation to my husband and family who have cheered me on along the way. Your love and support has spurred me on every step of the way and made my dream of a birthday publication date a reality.

ABOUT THE AUTHOR

Coach Laura Rubinstein is an award winning Social Media and Marketing Strategist, Certified Hypnotherapist, bestselling author and speaker. She is the President and co-founder of the Social Buzz Club (www.Social BuzzClub.com) and creator of the Savvy Social Media Success System.

Laura has optimized marketing plans and developed branding strategies for more than 1,000 businesses, celebrities, speakers and authors across the globe helping them create more profits and brand popularity. Her passion for working with entrepreneurs comes from her childhood and being raised in a household where her father and small business engineering professional worked from home.

She is the author of several marketing, personal development, and social media books including, *Appreciation Marketing, Your*

Guide to Growing Profitable Business Relationships. She is a contributing author to the Amazon bestselling books *Women Living Consciously* and *Journey to Joy*. She is the creator of the *Feminine Power Cards* (endorsed by John Gray, PhD author of *Men Are From Mars, Women Are From Venus*) which offer practical tools that allow people to make profound shifts in their relationships and professional life.

Laura, a Vanderbilt University School of Engineering alumnus, draws on an analytical nature, strong interpersonal skills, diverse business experience, team-oriented nature, and creative problem-solving abilities which make her coaching, branding, and marketing programs highly effective.

Her background and unique set of skills, knowledge and experience offer clients and audiences innovative strategies for building irresistible brands, buzz, and profitable and fulfilling relationships using social media.

Laura lives in San Diego with her beloved husband Kevin. They enjoy cooking and traveling while Laura is not writing and crafting social media strategies for her clients. Connect with Laura at www.TransformToday.com or on Twitter @CoachLaura.

www.ingramcontent.com/pod-product-compliance
Lightning Source LLC
Chambersburg PA
CBHW050109210326
41519CB00015BA/3895